P9-EJK-625

DNA

Pioneers

and

Their

Legacy

DNA Pioneers and Their Legacy

Ulf Lagerkvist

Yale University Press

New Haven and London

Publication supported with generous funding from the Wenner-Gren Center Foundation for Scientific Research, Stockholm, Sweden.

Parts of this book are adapted from a book previously published in Swedish: Ulf Lagerkvist, *Gener, molekyler, människor* (Stockholm: Brombergs Bokförlag, 1991).

Printed in the United States of America by Vail-Ballou Press.

A catalogue record for this book is available from the British Library.

Library of Congress Cataloging-in-Publication Data

Lagerkvist, Ulf.
DNA pioneers and their legacy / Ulf Lagerkvist.
p. cm.
"Parts of this book are adapted from a book previously published in Swedish: Ulf Lagerkvist, Gener, molekyler, människor (Stockholm: Brombergs Bokförlag, 1991)" — T. p. verso.
Includes bibliographical references and index.
ISBN 0-300-07184-1
1. DNA—Research—History.
2. Genetics—History. I. Title.
QP624.L34 1998
572.8'6'0072—dc21 97-37281

The paper in this book meets the guidelines for permanence and durability of the Committee on Production Guidelines for Book Longevity of the Council on Library Resources.

10 9 8 7 6 5 4 3 2 1

To the memory of

Einar Hammarsten,

who introduced me

to nucleic acids

Contents

Preface

When I first published in the nucleic acid field in the late 1940s, I never imagined the breakneck speed with which this subject would develop during the latter half of the twentieth century. At the time it seemed to be moving at a rather sedate pace, and the number of people active in the field was small compared to the innumerable scientists currently working on nucleic acids. To keep up with the literature then was still a manageable task, whereas the graduate student of today is swamped with information and cannot possibly digest it all. Knowledge of the pioneers—Friedrich Miescher, Albrecht Kossel, and others—is swept away and lost to sight in the constant struggle to keep abreast of the runaway development.

This issue might seem a small one compared to the pressing problems that science is faced with every day, but I believe that it is worth our consideration. The half-forgotten scientists who more than a century ago devoted their lives to this new field may seem to be only names in a textbook, of no real interest to the new generation of nucleic acid chemists and molecular biologists. Nevertheless, those who went before may help us to see our own efforts in a more revealing perspective. Further, they may teach us a useful lesson in humility and make us realize the truth of the old saying that if we can see as far as we do, it is only because we stand on the shoulders of giants.

What are my intentions with this book, and what kind of audience do I have in mind? My answer is that the book is intended for everyone who wants a popularized and simplified look at how it all started. In telling my story, I have tried to avoid being too scientifically ambitious, without being drawn into the trap of presenting mere entertaining chatter. I have sought to give a glimpse of the often colorful characters who helped to create this new science that has revolutionized biology. My own teaching in the biomedical field has convinced me that insight is something modern students very often lack. They have absorbed the scientific facts, but the historic background and a sense of the characters involved is missing. In this book I attempt to emphasize these aspects.

I have tried to present a story that, although biographic, is reasonably coherent. Perhaps it gives too much weight to the human beings involved and too little to the scientific facts, but the personal element is very important in science—and not only in terms of remarkable discoveries made by individual scientists. Equally important is the decisive influence a mentor can have on pupils—his or her ability to captivate young people and make them addicted to science. This process is at the very heart of the scientific tradition. In order to understand how it works, we must consider the personalities of the great scientists, not merely their research achievements.

My focus here is on DNA: its discovery, its role as genetic material, its structure and replication. I do not mean to lay out a history of molecular biology and molecular genetics, but rather to present some essays with a common theme. I hope to whet the appetite of my readers and tempt them to delve more deeply into the fascinating history of nucleic acids.

I am very grateful to Maja Samimi of the Friedrich Miescher-Institut for generously providing me with biographic material and for taking the trouble to answer my questions so fully.

Jacques Fresco, Arthur Kornberg, Joshua Lederberg, Göran Levan, and Tomas Lindahl were kind enough to read the manuscript and give helpful opinions. I am also much indebted to my former collaborators Per Elias and Tore Samuelsson for helping in preparing the figures and for comments on the text.

Vivian Wheeler has done a marvelous job of honing my English.

Finally, I gratefully acknowledge the generous financial support of the Wenner-Gren Center Foundation in the publication of this book.

[one] A Scientist's View of Science

The history of science is full of people who have worked under the most appalling conditions and have nevertheless made incredible discoveries. Further, in spite of a total lack of economic incentive they have managed to attract young people to this seemingly thankless pursuit. Those of us who are scientists and have become interested in scientific history are constantly faced with certain basic questions. What is it that makes scientific research so irresistible to some individuals that they dedicate their lives to it? What does it take to become a scientist? What qualities are important?

Who Are We and What Motivates Us?

Originality and independence have always been value-laden words in science, but at the same time there is something contradictory here. The importance of an intellectually stimulating milieu; the young scientist's need of a mentor, an idol whom he or she can attempt to emulate; the fact that some function well as junior members of a successful group but fail when they are on their own—how can all these factors be reconciled with the need for originality and an independent mind?

If we compare scientists to writers of fiction, it might seem that there are no similarities at all, except perhaps for the total dedication that both bring to their work. The ability of writers to use language as a medium to express ideas and feelings, the capacity to create living and interesting figures who can reach out and touch our hearts, would seem to be the result of unique personal qualifications. Scientists, in contrast, are the product of scientific tradition and milieu, and are therefore in a way interchangeable. Of course, writers of fiction are also part of a cultural and literary tradition, but one cannot help feeling that the work of writers is more uniquely their own than can ever be the case with scientific discoveries. In the fields of research with which I myself am reasonably well acquainted, it is impossible to think of a single great discovery that could not have been made by one scientist equally as well as another, perhaps a few years later. What remains, then, of all this talk about originality and independence in scientific research? To begin with, one must bear in mind that these terms have a somewhat different meaning in science than in literature and the fine arts. All scientific progress is the result of continuous development; for thousands of years scientists have added one stone after another to a constantly growing structure. One can always see how the latest insights relate to and are the logical consequence of an endless number of previous discoveries without which they would not have been possible. Pure chance might almost seem to determine who of the countless workers on the vast building site succeeds in placing the last stone.

This is not the whole truth, of course. There is certainly room for a unique, personal contribution also in science; this venerable edifice is not just an anthill crawling with innumerable, virtually identical worker ants. Nevertheless, there is some justification for claiming that science is the triumph of anonymity. This building, more lasting than the gilded monuments of princes, we have built with iron diligence and unfailing dedication—we highly competent, anonymous scientists without whose efforts scientific progress would have been impossible.

Scientific research may have its moments of existential loneliness, both in the hour of triumph and the hour of defeat, but it is in fact a fundamentally social function. In any case, this is true of medicine and the natural sciences. Scientists are a band of brothers, a statement that might sound a bit odd since we are often thought to be at loggerheads with one another. Nevertheless, the international scientific community is a reality. The feeling that science is without national boundaries, that what you are doing may be of the greatest importance for some like-minded disciple in Kyoto or Novosibirsk, is incredibly stimulating and constitutes one of the most important incentives to scientific work.

Like many mammal predators, scientists hunt in packs; the research group has its special hierarchy and social structure, much the same as the wolf pack. The leader of the group is ultimately responsible for directing the work toward goals that are at the same time important and attainable. He or she (for the scientific band of brothers nowadays admits a considerable number of sisters, I am happy to say) must see to it that the other members of the group can feel the confidence in their work that comes from competent leadership, without being inhibited in their ability to question theories or procedures and come up with new ideas and suggestions. It is a matter of the leader's having a grip that is firm without being stifling. This would seem simple enough, but in reality it can be quite difficult.

The junior members of the group, the graduate students, are the apprentices in science, and without their dedication and hard work the group could not function. The romantic picture of the scientist working alone in the laboratory is hardly the reality. Often one has to start out that way, but soon enough the experimental work becomes overwhelming and younger collaborators are essential. It is not only a question of workload: scientific thinking is a social function, an exchange of ideas between kindred spirits. It is a ball game where you must have other players to whom you can pass the ball and who will return it. Gifted and enthusiastic graduate students and postdoctoral fellows can give the kind of invaluable support to a senior scientist that is decisive to the ability to function intellectually. Without them the task would become impossible.

Of Risk Assessment and the Scientific Language

During the long history of science both the Catholic church and worldly authorities, and in later centuries also public opinion, have been much concerned with the risks involved in research. Originally it was a question of how the new knowledge might affect the religious beliefs of ordinary people, perhaps even endanger the salvation of their eternal souls. These days we have become more preoccupied with possible risks to the health and well-being of the public, rather than the prospects for eternal life. In any case, the assessment of risks, real or imaginary, has been a central issue in science since the days of Giordano Bruno and Galileo Galilei. A crucial part of this problem has always been our ability, or rather our inability, to communicate with the authorities and the public.

Once upon a time, when I was a small boy and children were still allowed to devour the fairytales of the brothers Grimm, I read the familiar story of the imp in the bottle. A man found a dusty bottle in a corner. Having examined it and found it to be worthless, he was going to throw it away when he discovered a small figure inside. It begged in a pitiful voice to be let out. When the man hesitated, the imp promised to grant him any wish he might have, if only he would release the imp from the bottle. It was a tempting proposition, of course; but when the man pulled the cork, the imp grew into a horrible monster once it had escaped its prison. The man was frightened out of his senses when he realized what he had done. It had been easy enough to release the imp, but it was obviously going to be an entirely different matter to put it back in again.

Many people regard the development of science and technology in this century with the same feelings that the man in the fairytale must have experienced after he drew the cork. We have let the genie out, and it has served us well. At the same time, we are afraid of what it may do now that we have let it loose—for we know that we cannot put it back. From this rational and understandable fear has grown a demand for absolute guarantees against all conceivable risks, however fanciful and far-fetched. What people forget is that no human activity can be entirely without risks. That is something we can never achieve; it is an illusion. Instead, we must try to assess the risks as correctly and precisely as possible and weigh them against the benefits. Self-evident and banal, you may say—but that does not make it any easier or any less important.

To take risks in order to achieve a coveted goal has always been part of human behavior; one might almost think of it as a heritage from our forebears who eked out a precarious livelihood among the great predators on the grassy savannas of Africa. Yet our attitudes toward different types of risks vary considerably. Risks with which we are well acquainted and that have become part of our daily life, like getting into the car and driving off to the job, are well tolerated and we accept them as inevitable. An important element in our ready acceptance is of course that we believe ourselves to be in control of the situation. Experience should have taught us differently, but it has not. Take the fear of flying. People who would not give a second thought to driving their car from New York to Boston break out in a cold sweat at the thought of sitting in an airplane, completely and helplessly dependent on the skill of a pilot about whom they know nothing from personal experience. The fact that he has been thoroughly trained to fly his machine as safely as possible and the fact that, statistically, flying is much safer than driving make no difference. It is the fear of the unknown that overwhelms the terror-stricken air passenger. Something of the same irrational fear can occasionally be seen in the layman's attitude toward science.

Can scientists be trusted, or are they a closed society that tries to keep out inquisitive and critical people? Are they really honest and open about possible risks, or do they keep dark secrets in order to preserve the privilege of free research? To those of us who are scientists, members of this international brotherhood, these questions may seem grotesque. Of course we can be trusted! But to the general public and to the media, our credibility is far from obvious. Instead, they tend to suspect a conspiracy to keep them in the dark. What are scientists really up to in their laboratories and what is secretly growing in their bottles? In the realm of entertainment this theme has been beloved ever since Frankenstein, and unfortunately something of the same attitude prevails in the world of reality.

Are we really so secretive? Everyone who has any knowledge of scientists knows that they love to talk about their research. The problem is finding someone willing to listen. Having exhausted all other audiences, only their scientific colleagues remain. So much for scientists as a secret society that wants to keep out journalists and concerned citizens! Yet if we are all that eager to communicate, how did the dark legend of our mysteriousness arise? It seems to have something to do with scientific language; people do not understand what we are saying. We are victims of the Babylonic curse, the punishment that Jehovah devised for the arrogant descendants of Noah when they built the Tower of Babel to reach all the way to the sky.

Even the most successful translation of scientific facts into everyday words and concepts is to some extent a deformation and oversimplification of a much more subtle reality. All translations from one language to another are more or less unsatisfactory. There are even people who maintain that all attempts to popularize science are doomed to failure and must be rejected as violations of the truth. As far as I can see, this fundamentalist attitude is based on an erroneous conclusion: since popular science necessarily represents a simplification, it is always misleading and has no function in the communication between the specialist and those outside that field of research. This conclusion may seem attractively simple and straightforward, as the truth ought to, but it is all wrong.

For practical purposes we can distinguish between the specialist's knowledge, which gives the ability to formulate and solve problems in that specific area of research, and call such knowledge "operative"; let us denote all popularized knowledge as "journalistic." The fundamentalists claim that journalistic knowledge is useless and perhaps even harmful. However, the fact is that journalistic knowledge is very useful even to a scientist; this is the way we communicate with scientists in other fields.

Let me give an example from my own research. My knowledge of the sophisticated physical methods that can be used to characterize nucleic acids and proteins in terms of their three-dimensional structure is entirely journalistic. I know the principles of, for instance, X-ray crystallography, but my knowledge is definitely not operative. Nevertheless, on the basis of my journalistic knowledge I can communicate with my friends the crystallographers. I know what they can do to solve my problems, and what I must do to make this possible for them. The claim that journalistic knowledge is useless can therefore be refuted on the basis of our experience within science itself.

Why then do we so often fail in our attempts to communicate with the media and the public? One might say that certain conditions must be met in order for this communication to succeed. As far as popular science is concerned, a tacit agreement exists between the author and the reader. The author undertakes not to use foreign words and terminologies without first explaining them, and not to assume previous knowledge that ordinary people cannot reasonably be expected to possess. The reader, on the other hand, promises to make a sustained effort. The question, of course, is how much of an effort can realistically be expected in order to master a popular scientific text.

Let us assume that someone in anticipation of retirement decides to learn how to play chess. He (or she) has heard of this ancient game as a stimulating pastime that for many chess addicts has become a passion and a way of life. Otherwise he knows nothing about chess, and he therefore purchases an elementary book on the subject. He obviously realizes that he is required to make an effort to learn how the pieces move, what opening gambit to use, and so on—but he is prepared to make this effort. This is exactly the attitude a writer of popular science must hope for in his reader. Feelings of joy and excitement must be experienced by the reader, otherwise the battle is lost before it is joined. To paraphrase an old proverb, one can lead a reader to the well of knowledge, but she must drink from it herself; no one can do it for her.

To return to the question of research and risks, let us tentatively conclude that it is possible to make reasonable assessments of both benefits and putative risks that might result from our research, and that it is possible to communicate our views in an intelligible way to the public and the media. In the ensuing discussion, a major question will always be, How much risk would it be justifiable to take? Certainly we must be prepared to accept some risks. After all, scientific research is a venture into the unknown and absolute guarantees are not possible. Furthermore, it is a trivial but important truth that we may encounter risks also by abstaining from research projects out of fear of unknown dangers.

To illustrate, let me draw a parallel to the dawn of bacteriology—a time that might with some justification be called the heroic period of medical research. Suppose that the authorities in those days had turned on Louis Pasteur and Robert Koch and forbidden bacteriological research because those great scientists could not guarantee that their work was completely safe and free of risk. When we look back after more than a century on what the pioneers of bacteriology achieved when they laid the foundations of this new science, we can agree that it would have been unjustifiable and even dangerous to prohibit bacteriological research. We would have been deprived of all the methods for diagnosis and treatment of contagious diseases that have saved countless lives since the days of Pasteur and Koch. We cannot deny that they took certain risks, but the results assuredly justified their decisions.

Of course it is necessary to control the practical applications of new knowledge, and in certain cases to forbid projects that seem dangerous or unethical. But one cannot prohibit the search for knowledge, regardless of how well intended such a ban might be. When the Catholic church almost four hundred years ago burned Giordano Bruno at the stake and later threatened Galileo with the same fate, its commendable intention was to prevent people from listening to heresies and thereby endanger their prospects of salvation and eternal life. The church had the noblest of motives, it would seem, but nevertheless made a catastrophic error that has haunted it ever since.

Intellectual curiosity is a characteristic of human beings, and during the long history of scientific research it has never been possible for the powers-that-be to stifle it entirely. Our insatiable curiosity is an inheritance that will always be with us. In that sense at least, science can face the future with a certain amount of confidence.

[two] A Young Science and Its Roots

Medicine is as old as the human race. From time immemorial humankind has tried to cure diseases and heal injuries. The basic medical sciences—anatomy, physiology, pharmacology—can be traced back to the first efforts in the golden age of Greece to find a scientific basis for medicine. The theories of Hippocrates about the supreme importance of the four body fluids (blood, phlegm, yellow and black bile) and their relative proportions may today seem far-fetched and fantastic, but by their rejection of religious and magical explanations of medical problems they represent a decisive step toward a scientific medicine.

The rational approach of Hippocratic surgery to the treatment of fractures and luxations is clearly based on an understanding of the anatomy and function of the skeleton, and to a certain extent also of ligaments and tendons. The prolific writings of Galen in the second century A.D. include the results of systematic dissections of animals (for religious reasons, human bodies were not available) but also important discoveries in neurophysiology based on experiments with living animals. Pharmacology too is extensively treated by

Galen, and here his writings are part of a long line of ancient pharmacopoeias that have come down to us from both Egyptian and Chinese medicine, although it is doubtful that Galen was much aware of these predecessors.

Today biochemistry is the dominant basic science of medicine in the sense that it outstrips all the others in terms of the number of publications per year. Nevertheless, it is something of a newcomer, at least compared to such veterans as anatomy and physiology. In the sections that follow I shall trace the roots of biochemistry from the days of Paracelsus to the end of the nineteenth century.

Prophets and System Builders

Deep is the well of the past . . . . Thus begins the story of Jacob in Thomas Mann's famous series of novels with motifs from the Old Testament. Let us look into the well and see if we can discern anything in its dark depths—a face, perhaps?

Philippus Aureolus Theophrastus Bombastus von Hohenheim, who called himself Paracelsus to emphasize that he was the equal—perhaps even the superior—of the classic authorities Hippocrates, Celsus, and Galen, is one of the true iconoclasts in the history of medicine. He was born in Switzerland in 1493, the son of a Swabian physician. When he decided to follow in his father's footsteps like so many doctors' sons before him, that was the only conformistic action of this thoroughly unconventional and revolutionary character. In a picture from 1538, he is portrayed as a prematurely aged, balding man with an arrogant profile and a thin-lipped, bitter mouth. The face of a man who easily made enemies, one would think.

Physician, alchemist, visionary, and rebel, Paracelsus was entirely medieval in his superstition and lack of critical scientific judgment. At the same time, he was remarkably modern in his rejection of all conventional medical wisdom that had been inherited from antiquity without ever being questioned. In his writings he appears as a messianic superman who roundly denounces contemporary medicine and in a furious outburst prophesies that dogs one day shall pee on all the learned doctors at the Sorbonne and other universities who had tried to thwart him and refused to acknowledge his greatness. When for a short time he was a professor of medicine at the University of Basel, he lectured in German instead of Latin, the traditional language of learning; and as an introduction to his lectures he solemnly burned the books of Galen and the Arab physician Avicenna, to demonstrate his complete repudiation of the paladins of medicine.

Instead of the time-honored teachings of Hippocrates, with his emphasis on the body humors and their relative proportions as the basis for all medical thinking, Paracelsus put forward an all-embracing theory of his own. Sickness and health were explained in terms of three chemical principles: *sulfur*, which signified combustibility; *mercury*, which symbolized fluidity and changeability; and *salt*, which stood for solidity and stability. The delicate balance of these chemical principles in the body was essential for its normal function, and disturbances in the balance caused disease.

Undoubtedly, the theory was influenced by the interest of Paracelsus in alchemy—he was fascinated by the possibility of transforming base metals such as mercury and lead into gold (a prospect dear also to the hearts of the kings and princes of those days, with their empty coffers and unquenchable thirst for gold). Alchemy embodied all the knowledge of inorganic chemistry of the time, and it is logical that Paracelsus was the first to introduce the salts of heavy metals into the pharmacopoeia. This addition to the therapeutic arsenal became very popular and must have caused untold suffering over the centuries.

It is easy to dismiss Paracelsus as a dangerous lunatic, and the majority of his medical colleagues undoubtedly would have subscribed to this view. He was surely one of the most-hated men of his profession. Yet his fantastic theories can be seen as the first attempt to explain physiological phenomena in chemical terms. He was an early advocate of the importance of the natural sciences in medical training, always emphasizing that physicians must have a background in science. Chemistry was obviously important, but so was astronomy with its sister science astrology, since the stars influenced the human body and its functions. Thus, the teachings of Paracelsus are a remarkable and enigmatic tissue of the crudest superstition alternating with visionary glimpses of what would one day be called molecular medicine. When he died at the early age of forty-eight (because of a dissolute life, according to his detractors), he left an important legacy to posterity that would influence medicine for generations.

Paracelsus was the founder of iatrochemistry, a school of thought that breaks with the Hippocratic tradition and seeks to rationalize medical problems in chemical terms. Perhaps the best-known representative of this school was the Flemish nobleman Johann Baptista van Helmont (1577–1644). It is difficult to imagine two more disparate characters than Paracelsus and van Helmont. Paracelsus, an irascible gnome who even in the far-from-fastidious sixteenth century was uncouth and untidy in his personal habits, would heap the coarsest invectives of vulgar language on his adversaries at the leading medical faculties. Where Paracelsus shouted abuse at the top of his voice, the aristocratic van Helmont was quietly sarcastic as he criticized the absurdities of Hippocratic medicine.

Nevertheless, van Helmont was very effective when he ridiculed Hippocrates' obsession with the role of the dangerous phlegm, the bogeyman of classical medicine, in causing diseases of the lung, for instance. As is so often the case, van Helmont was more successful in his criticism than he was in the production of alternative explanations. His collected writings, *Ortus Medicinae*, are full of unsupported speculations; in that respect he was a true follower of Paracelsus. Still, he made a major chemical discovery when he described the formation of "gas sylvestre" (carbon dioxide) as the result of both the burning of charcoal and the fermentation of must. He also coined the term "gas" as distinct from "air" and "water vapor." Johann van Helmont was fascinated by ferments and fermentation, and theorized that all processes in the organism were caused by ferments that converted food into living flesh in six steps.

These ideas were taken up and elaborated by the German physician Franz de le Boë, called Sylvius (1614–1672), who unlike van Helmont practiced medicine with great success and also trained a large number of medical students. The salient points in Sylvius' influential teaching were that fermentation gave rise to both acid and alkaline end products, and that a lack of balance between acid and base in the body was the cause of disease. This concept of ferments and their central role in the living organism became very popular in the iatrochemical school and, superficially at least, it might seem that van Helmont and his followers were on the right track. However, as we shall see, the term "ferment" meant different things to different authors and is not the equivalent of the modern word "enzyme." It is dangerous, therefore, to overinterpret the ideas of the iatrochemical school and ascribe an insight to its supporters that they did not possess.

The great English chemist Robert Boyle (1627–1691) was no admirer of Paracelsus. In his book *The Sceptical Chemist*, published in 1661, he contemptuously rejected the notion that "sulfur," "mercury," and "salt" were the fundamental principles of what we today would call biochemistry. Boyle performed physiological experiments and probably realized the connection between respiration and combustion.

In any case, this was certainly true of his younger colleague in the Royal Society, John Mayow (c. 1640–1679). Mayow had a checkered career, obtaining a degree in law before he switched to medicine and became interested in the problem of respiration and combustion. His description of the respiratory apparatus in animals and its action during respiration is remarkably accurate. He understood that air contained something that he called "spiritus nitroaereus," which was necessary for combustion. In a series of experiments he showed that spiritus nitroaereus was consumed not only by a burning candle enclosed in a bell jar, but also by a mouse breathing in such a jar. When the spiritus nitroaereus was used up, the mouse died and the candle went out. Mayow also realized that spiritus nitroaereus was taken up by the blood when it passed through the lungs, and that this was the reason for the characteristic change in color from the dark red of venous blood to the light red of arterial blood.

The originality and importance of Mayow's contributions, compared with those of, say, Robert Boyle or Robert Hook, have been the subject of some debate. Nevertheless, one might think that, with Mayow's discovery of what we now call oxygen, an accurate theory of combustion and respiration would have been close at hand. Instead, there occurred one of those bizarre episodes in the history of science when an erroneous but generally embraced theory completely misleads generations of scientists.

Since the days of antiquity, fire had been seen as something substantial, as one of the four elements (together with earth, water, and air). The notion that fire is present in latent form in any combustible material probably dates to the time when man first discovered what a useful servant it could be. Coal and sulfur, for instance, were thought of as containing a "substance" that was liberated during combustion and could then be observed as a flame. At the end of the seventeenth century, the German physician Georg Ernst Stahl used the term " phlogiston" to designate this latent form of fire.

In those days enormous progress was being made in the sciences of mathematics, chemistry, and physics by men such as Robert Boyle, Isaac Newton, and Gottfried Leibniz. But in medicine it was a time of barren speculations and the building of grandiose pseudophilosophical systems. Like the other leading physicians of the period, Stahl had his own system that supposedly could explain everything; but like the other systems, Stahl's was short-lived and had no lasting influence. The same unfortunately was not true of his brainchild, phlogiston. For a century it would warp the thinking about combustion and respiration, and prevent scientists from drawing the correct conclusions from Mayow's discovery of spiritus nitroaereus.

Although Stahl had studied the formation of metal oxides, he explained the phenomenon not as the uptake of oxygen from the air (as we would describe it) but as the loss of phlogiston. The phlogiston theory takes us into a kind of upside-down world, where combustion is seen as a process in which the burning material loses its latent fire (phlogiston) to the surrounding air, which consequently becomes enriched with phlogiston.

When the Swedish apothecary and chemist Carl Wilhelm Scheele in 1772 produced pure oxygen by heating metal oxides, he was still completely captivated by the phlogiston theory. The English chemist Joseph Priestley, who independently made the same discovery two years later, actually called the new substance dephlogisticated air. This was quite logical according to the theory: the more oxygen in the air the less phlogiston, and vice versa. This was how chemists viewed the problem of combustion until Lavoisier appeared.

Antoine-Laurent Lavoisier was born in Paris in 1743, the son of a high official of the Paris Parlement—which, incidentally, was not a parliament at all in our sense of the word, but a kind of high court. He came from a wealthy family and had a fine education. Eventually, his influential father got him a well-paid sinecure in the royal tax bureaucracy and also left him a title of nobility—a dangerous bequest, as it would turn out. As a result, Lavoisier could pursue his interest in chemistry and physiology without any economic worries. As early as 1768 he was elected to the French Academy of Sciences, and in 1772 he could report to the academy that when a combustible substance such as sulfur burned in air, something was taken from the air. Lavoisier realized that this something must be the new substance discovered by Scheele and Priestly. He renamed it "oxygen" and explained the process of combustion not as the loss of phlogiston, but as the uptake of oxygen that combined with the burning material. Thus, Mayow's spiritus nitroaereus was resurrected and Stahl's unfortunate phlogiston theory, which had caused so much confusion, was finally discarded.

Lavoisier, who also studied respiration, confirmed Mayow's

view that in the lungs oxygen is taken up by the blood and that this is the fundamental phenomenon in respiration. He realized that a warm-blooded animal produces heat from the combustion of food, with the formation of carbon dioxide and the consumption of oxygen. Consequently, this process could be compared to the burning of charcoal in air. Using a calorimeter, he demonstrated that the heat given off by an animal, which produces a certain amount of carbon dioxide during the experiment, is roughly equivalent to the heat generated by the burning of a piece of charcoal to give the same amount of carbon dioxide. His arguments may have been oversimplified and did not take into account a number of complicating factors unknown to him, but he nevertheless made the first attempt to outline the basic principles of energy metabolism. If anyone deserves to be named the father of biochemistry, it must surely be Lavoisier.

Apparently unaware of the political convulsions to come, Lavoisier led a happy family life. Nothing could seem more idyllic and remote from the evils of this world than David's famous painting of Lavoisier in his laboratory together with his wife, who like a proud mother with her prodigy child puts a hand protectingly on his shoulder. But in 1794 the French Revolution had entered into its last, self-destructive stage during the Terror. Because of his social standing and the position he occupied, Lavoisier was in the end drawn into its fatal whirls. On May 8, 1794, he was executed in the Place de la Revolution with the help of Dr. Guillotin's convenient apparatus for the rapid and painless disposal of the enemies of liberty. A number of bestial crimes were committed in the name of liberty and equality during the French Revolution, but none, in terms of simple, monumental idiocy, could compare with this act. The next day Joseph-Louis Lagrange, the renowned mathematician, sadly remarked, "It required only a moment to sever that head, and perhaps a century will not suffice to produce another like it."

The Enigmatic Ferment

Probably no single chemical reaction can compare with the formation of alcohol by the fermentation of sugar, in terms of the enormous interest it has generated among both laymen and scientists. Fermentation to produce alcohol required yeast, which early on was designated as "ferment." However, in the heyday of alchemy in the thirteenth to fifteenth centuries, the term "fermentation" was used very loosely to indicate also processes involving inorganic substances like metals and their imagined transformations, for instance in the making of gold. In the same vein, "ferment" could mean all sorts of magic elixirs and potions.

During the sober and rational eighteenth century most of these notions were discarded. In his *Elements of Chemistry* the famous Dutch physician Hermann Boerhaave restricted fermentation to mean solely a process in which an organic substance was decomposed to give either an alcohol or an acid. As ferment he defined anything that could give rise to this process (presumably what we today call microorganisms). In 1680

Boerhaave's countryman Anthonie van Leeuwenhoek had examined yeast in a simple, remarkably high resolution microscope of his own design and making, and had seen small spherical particles. Not that Boerhaave's book mentioned Leeuwenhoek and his yeast cells: it would take about one hundred fifty years before the importance of the discovery was realized.

Carbon dioxide had already been discovered by van Helmont, but it was not until the middle of the eighteenth century that the Scotch chemist Joseph Black could definitively demonstrate that the gas, which had been seen from time immemorial as a foam on the surface of the fermenting brew, was indeed the same gas he had previously produced by heating magnesium carbonate. It was a decisive step forward in our understanding of the chemistry of alcohol fermentation. From his studies of *la fermentation vineuse* Lavoisier drew the prophetic conclusion that the fermented sugar was separated into two parts, one that was oxidized to carbon dioxide and one that lost oxygen and gave rise to alcohol. It would seem to be a short step indeed from this insight to a correct concept of alcohol fermentation, but instead it would take a century of intense research and the most magnificent scientific quarrels before the goal was eventually reached.

The real stumbling block was the true nature of the ferment. For us it is natural to associate this term with the modern word "enzyme," but as we have seen, this was not how the term had been used in the past. To illustrate the confusion that prevailed as recently as two hundred years ago, one example may suffice.

In the year 1800 the French Academy of Sciences, or rather its successor during the Revolution, announced an essay competition: it wanted an answer to the question of what distinguished the ferment from the substance being fermented. The prize for the best answer was a one-kilogram medal of pure gold. Unfortunately the competition was later canceled and the generous prize offer withdrawn for lack of funds. The question posed might seem somewhat bizarre, but it becomes understandable if one realizes that the scientific community had long discussed the possibility that a sugar molecule being fermented to alcohol and carbon dioxide could transfer this capability to other sugar molecules—infect them with fermentation, as it were—by way of something referred to as "molecular vibrations." Even in the middle of the nineteenth century the renowned German chemist Justus von Liebig adhered to such ideas.

Against this background it is easy to see what a huge step forward occurred when the French engineer Charles Cagniard-Latour, and independently of him the German biologists Theodor Schwann and Friedrich Kützing, in 1837 described the yeast cell (Leeuwenhoek's discovery had long been forgotten) and clearly showed that it was necessary for alcohol fermentation. The yeast cell was in fact the enigmatic "ferment," and one could forget such nonsense as "molecular vibrations." A huge step forward indeed, the only trouble being that in their enthusiasm the supporters of this cellular theory of fermentation allied themselves with a romantic natural philosophy and its notions of a mysterious vital force that could not be explained in chemical terms.

This vitalism, as it was called, counted the French physician Théophile de Bordeu as one of its founding fathers. Something of a mystic, he taught that the functions of the human body depended on processes that were inseparable from the living organism and could not be reproduced outside it. Vitalism gained enthusiastic support in Germany, where the leading romantic philosopher was now Friedrich Wilhelm Schelling (1775–1854), an outstanding pupil of Kant. At the early age of twenty-two the precocious Schelling published his book *Ideen zu einer Philosophie der Natur*: It became a canon of natural philosophy and was immensly influential at the time. Schelling considered Nature to be possessed of a soul; in his opinion even inanimate material showed signs of life, as demonstrated by such phenomena as electricity and magnetism. His teachings aroused enormous, almost religious enthusiasm, not least among German physicians, where Schelling had an ardent prophet and interpreter in Dietrich Kieser. Particularly appealing to doctors, of course, was the fact that Schelling considered medicine to be the foremost of all sciences and the one closest to the Divine Being.

A central thought in Schelling's natural philosophy, further elaborated by the faithful Kieser in his *System der Medizin*, was the polarity that characterized everything in the universe. Life was seen as oscillating between positive and negative poles, between the positive sun and the negative earth. The male character was mainly influenced by the sun, whereas the female nature was more earthly. Disease was considered to be caused by a disturbance of the natural polarity, but could alternatively be explained as a fall from a higher to a lower level in the hierarchical structure of creation, where man was at the top of the biological ladder. This rank nonsense dominated theoretical medicine in large parts of Europe in the early nineteenth century, and vitalism played a major role in the acrimonious quarrel about ferment and fermentation that continued until the end of the century.

Vitalists rejected the idea that organic compounds found in the cell could be synthesized in the test tube of the chemist. The formation of such compounds depended on a vital force in the organism. Consequently, a process such as the fermentation of sugar to alcohol was possible only in the living yeast cell. The leading advocate of these ideas was none other than Louis Pasteur (1822–1895), known as the father of microbiology. Exerting all his experimental genius and dialectic brilliance (Pasteur's ability to shatter his opponents in debate was legendary), he contended that fermentation—and here he included a number of reactions other than alcohol fermentation—was absolutely dependent on the presence of living microorganisms and could not take place outside the cell. That was what his data seemed to indicate, but certain facts did not fit into the simple and unambiguous picture that Pasteur painted. Thus, his opponents under their leader, Justus von Liebig (1803–1873), could point to the existence of a number of soluble ferments, as they were called, that had been obtained as cell-free extracts. These soluble ferments catalyzed just the kind of reactions that according to the vitalists should be the prerogative of living cells, for instance the cleaving of sucrose to fructose and glucose.

Pasteur's dialectic talents proved up to the challenge; he had an answer for everything. He introduced a new term, *fermentation proprement dit* (fermentation in the proper sense of the word), and defined it in such a way that he nimbly excluded all reactions that did not require the presence of living organisms. That left Liebig and his supporters with long faces, swearing under their breath about circular arguments.

Pasteur's relations with his German critics were hardly improved when the war between France and Prussia broke out in 1870. The whole glittering Second Empire, with its flaking gilt over a rotten core, came tumbling down like a house of cards before the onslaught of the Prussian military machine, egged on by that brutal political genius Bismarck. Pasteur was a great French patriot, and his attacks on Liebig became even harsher than before. Liebig, a romantic and sensitive soul, died in 1873. Even if his conflict with Pasteur did not actually shorten his life, it certainly did nothing to brighten his last years. With all his romantic exaggerations and his molecular vibrations, Liebig was ultimately proved right in a way. To be fair, he had been supported by French scientists also. The well-known chemist Marcelin Berthelot in 1860 categorically rejected "life" as a satisfactory explanation of such phenomena as fermentation. He argued that one should instead use chemical methods and look for what we today would call molecular explanations of biological processes.

The soluble ferments that Pasteur wanted at any cost to keep out of the discussion of the true nature of fermentation undoubtedly represented a momentous discovery, just as Liebig and others had maintained. Those ferments were of course enzymes, although at the time that term had not been introduced. Enzymes are biological catalysts that act in practically all chemical reactions in the cell. As early as 1835 the Swedish chemist Jöns Jakob Berzelius (1779–1848) had defined catalysis and linked it to the mysterious ferment. Berzelius was undoubtedly the leading chemist in the first half of the nineteenth century and with his *Readings in Animal Chemistry*, published in 1806, he was also one of the pioneers of biochemistry.

Berzelius attracted a number of pupils and younger collaborators to his laboratory at the Karolinska Institute, among them the German Friedrich Wöhler, who became a close friend of both Berzelius and Liebig. Wöhler was the first chemist to succeed in synthesizing an organic substance, urea, in the laboratory. This might well have been a death blow to the vitalists, but they wriggled out of the dilemma with the sophistic argument that urea was after all only an excretion product.

Wöhler and his longtime friend Liebig had obtained the bitter-tasting substance amygdalin in pure crystalline form from bitter almond, and in 1837 they succeeded in extracting from the same source a soluble ferment. They considered it to be "albuminoid" in nature (what we today would call a protein) and showed that it could split amygdalin. The number of soluble ferments increased steadily until at the end of the century about twenty were known. When the German biochemist Willy Kühne suggested in 1878 that these ferments should be called "enzymes," it was already obvious that they were indeed the biological catalysts predicted by Berzelius. Still, there was considerable uncertainty about their chemical nature: were they really proteins, as Liebig and Wöhler had believed?

With the very primitive methods of analysis available at the time, it was often impossible to demonstrate the presence of protein in extracts with enzymatic activity. Many scientists therefore refused to believe that enzymes were proteins. However, the pendulum swung in that direction when the brilliant German organic chemist Emil Fischer became interested in enzymes. He advanced the seminal hypothesis that the substrate fitted the enzyme as a key fits a lock, and he had no doubts whatever about the chemical nature of enzymes. They were proteins. With Fischer's enormous authority behind it, this view dominated biochemistry in the twentieth century.

As always there were dissidents, chief among them the German organic chemist Richard Willstätter (1872–1942). When Willstätter relatively late in life took up the study of enzymes, he was already world famous and had received a Nobel Prize for his work on the structure of chlorophyll and other natural pigments. He introduced a new method for the purification of enzymes, based on their adsorption and subsequent elution from inorganic gels. This method was so efficient that Willstätter's highly purified enzyme solutions seemed devoid of proteins, at least when analyzed with the methods available in the 1920s. The situation seemed to be back where it had been before Fischer pronounced his verdict on the nature of enzymes, and Willstätter became convinced that the great chemist had been wrong after all. He suggested instead that enzyme activity was often found associated with proteins simply because the true enzymes, whose chemical nature was unknown, had a tendency to bind to proteins, much the same as they bound to gels in his own purification procedures.

Because of the distinguished name he had made for himself as a chemist, Willstätter's opinion carried much weight. It would take the crystallization of urease by James Sumner in 1926 and the work on crystalline pepsin in the early thirties by John Northrop to free the biochemical community from the effects of Willstätter's mistake. Willstätter himself remained convinced of the nonprotein nature of enzymes until his death in Switzerland in 1942 as a refugee from the Nazis. No doubt he would have been immensely gratified if he could have had a look at biochemistry in 1983, when two young Americans, Sidney Altman and Thomas Cech, independently showed that certain RNA molecules can function as enzymes.

What, then, became of the powerful struggle between vitalists and chemists? It was decided in favor of the chemists in 1897, two years after the death of the vitalist chieftain Louis Pasteur, when Eduard Buchner, a pupil of the famous German chemist Adolf von Baeyer, demonstrated fermentation of sugar to alcohol in a cell-free extract of yeast. His old teacher von Baeyer never believed that Buchner would amount to much; when he learned of these exciting results, he reportedly said, prophetically but somewhat unkindly, "This will make him famous, in spite of the fact that he has no talent as a chemist!"

Buchner had obtained his yeast extract by subjecting the cells to high pressure. The truth of the matter is that he did it to help his brother Hans, a bacteriologist who had some exotic ideas about using the extract for medicinal purposes. To preserve the extract, Eduard added sucrose. He must have been enormously surprised when all of a sudden it started to ferment. With this somewhat fortuitous discovery, vitalism was finally laid to rest. It has never been heard of since, while Eduard Buchner went on to become world famous and got his Nobel Prize, untalented chemist that he was.

From Albuminoid Substances to Proteins

The observations that the content of the egg solidifies when heated, that milk curdles when acidified, and that blood coagulates outside the body must be very ancient indeed. Even in the beginning of the nineteenth century properties such as these were the principal characteristics used to define a group of biological substances commonly referred to as albuminoid. For practical reasons attention was focused first on easily accessible albuminoid substances such as egg albumin, the casein of milk that precipitated on acidification, and the thread-like, insoluble fibrin that the Hippocratic physicians had already observed in the clotting blood (and that played a major role in their theory of the body humors).

At the end of the eighteenth century, through the work of Claude Louis Berthollet it became clear that albuminoid substances contained considerable amounts of nitrogen, while Scheele demonstrated the presence of sulfur in egg white. When it was realized that albuminoid substances could be found in abundance in both animal tissue and plants, terms like "albumin," "casein," and "fibrin" were used also to designate the substances found in plants. It would take a long time indeed before the extreme diversity of the albuminoid substances was fully realized.

Back in 1838 the Dutch chemist Gerardus Mulder had introduced the term "protein," as opposed to "albuminoid substances." This name, derived from the Greek word *proteios* (chief), was suggested to Mulder by Berzelius, to signify that protein was the chief constituent of living organisms. It would, however, take until the end of the century before the term "protein" was in full use.

In what follows I will for convenience use the word "protein" even when the author in question adhered to the older terminology. Mulder did not distinguish clearly among different proteins and undoubtedly believed that protein was the source of all other substances found in the organism. This view was accepted by prominent persons such as Liebig, whereas the crusty Berzelius was critical and took a dim view of such easy generalizations. In spite of this criticism, Mulder's thoughts prevailed during the 1840s. By midcentury the enormous diversity of proteins was beginning to be realized, and the concept of protein as a homogeneous entity, the source of everything else in the cell, was discarded.

It was observed early on that proteins could be cleaved by acid or alkaline hydrolysis and would then give rise to products that were later called amino acids. Leucine was isolated as early as 1819 and glycine, the simplest of all amino acids, was obtained in 1820. In 1846 the first aromatic amino acid, tyrosine, was crystallized by Liebig. As the years went by, more and more amino acids were isolated from proteins—mainly after acid hydrolysis, which proved to be least damaging to the amino acids. At the turn of the century, a dozen amino acids had been obtained in pure form, but not until 1936 was the list of standard amino acids in protein brought to twenty with the discovery of threonine.

A significant breakthrough in our understanding of amino acid and peptide chemistry came as a result of Emil Fischer's work during the last two decades of his life. Bringing all his synthetic skill and immense knowledge of organic chemistry to bear on this problem, in 1907 he could point to the synthesis not only of a number of amino acids but also of peptides containing more than a dozen amino acid residues. He called them polypeptides and considered that they approached the size of natural polypeptides. He was loath to accept the staggering molecular weights of 12,000 to 15,000 assumed for some proteins by other authors. A long way remained before organic chemists would realize the true dimensions of protein molecules.

When, after 1850, the oversimplified views of Mulder gave way to a more realistic appreciation of the immense diversity in the protein world, it was to a large extent the result of the introduction of new methods for the fractionation of proteins. It was discovered that proteins might be separated from one another by taking advantage of their differential solubility in salt solutions of varying concentrations. Prosper-Sylvain Denis in particular made extensive use of the "salting out" method and showed that serum proteins would be fractionated in this way. Salt fractionation became one of the most widely used methods for separating proteins, and toward the end of the nineteenth century it was obvious that the cell contained a vast number of different proteins.

At this point, properties such as osmotic pressure and the effect on freezing point of different dissolved substances were being used to determine molecular weights. When these methods were applied to proteins, the results were astonishing. For instance, in 1891 a molecular weight of 14,000 was obtained for egg albumin, and in 1905 the even more unbelievable value of 48,000 was reported for hemoglobin. It is easy to understand that Emil Fischer was shocked and incredulous, but little by little the exciting world of biological macromolecules was revealed to the fascinated biochemical community. A new era was beginning. Further, in the second half of the nineteenth century a new group of biological macromolecules had been discovered by an obscure young Swiss physician. It would take a long while before their importance was fully realized.

[three] A Modest Pioneer

In the year 1844 the city of Basel was still suffering economically from the separation of the town from the surrounding rural half-canton of Basel-Land, which had revolted against the domination of Basel in 1831 and obtained its independent status in 1833. Working conditions were far from ideal at the university, founded in 1460 by Pope Pius II. Yet it could boast such luminaries as Erasmus among its long row of famous past scholars.

Friedrich Miescher, a young professor at the medical faculty, had every reason to be concerned about his future in Basel. He had studied in Berlin under the well-known physiologist Johannes Müller, who in his youth had been an ardent follower of Schelling and throughout his life remained vaguely vitalistic in his general scientific outlook. Nevertheless, Müller stood for a return to common sense in German physiology. An inspiring teacher, he founded an outstanding school that counted among its pupils such dignitaries as the father of cellular pathology, Rudolf Virchow, and the renowned physiologist and theoretical physicist Hermann von Helmholtz.

While in Berlin, Miescher had submitted a thesis on the development and pathology of bone tissue that was considered so excellent that a year later, in 1837, he was appointed professor of physiology and general pathology in Basel. The field that he was supposed to cover seems rather extensive in our eyes, but it was an era long before the intense specialization of modern medicine had set in. On the contrary, there was a long tradition at European universities of professorships that encompassed vast areas of medicine; the famous Swiss physiologist Albrecht von Haller had held a chair at Göttingen in the previous century that included anatomy, surgery, medicine, and botany.

If Friedrich Miescher was not unduly worried about the extent of his teaching and research commitments, he was extremely concerned about the economic difficulties of the university, the more so as he was newly married. His wife, Antonie His, was the granddaughter of the well-known Basel politician Peter Ochs. Miescher himself came of peasant stock in the little village Walkringen in the Emmental, and his marriage into the Ochs family should have represented an improvement socially for him. However, there was a problem.

Peter Ochs had been infected with the ideas of the French Revolution and was a great admirer and follower of Napoleon. During the period of French domination in modern Swiss history known as the Helvetian Republic, Ochs was extremely influential and in fact drew up the constitution for the republic. When the new republic became immersed first in the revolutionary wars and later in the Napoleonic wars, the population suffered severely as the country became a battlefield for

Austrian, Russian, and French armies. As French oppression increased and Swiss soldiers were conscripted by Napoleon for his campaign against Russia in 1812, popular support for the Helvetian Republic decreased dramatically. Under the Restoration after the Congress of Vienna Peter Ochs was, in fact, widely regarded as a traitor. The very name Ochs had a bad sound to it in those days, so Antonie's father decided to take his French grandmother's name, His.

A Nephew and His Uncle

On August 13, 1844, Antonie presented her husband with a son, and the happy parents decided to give their firstborn the same Christian name as that of his father, Friedrich. (This decision creates something of a problem for the biographer; to distinguish between father and son in what follows, the former will be referred to as Friedrich Miescher Sr. and the latter will be called simply Friedrich Miescher.) With his additional responsibility as a father and head of family, Miescher Sr. understandably found his position in Basel too insecure and later in the year moved to Bern. Taking up a position as professor at the university, he remained in Bern for six years. In 1850 he was back in Basel, where he now occupied the chair in pathology. Conditions at his former university had improved with the growth of the city both economically and in terms of population, particularly after the Sonderbund wars had ended in 1848 with the defeat of the reactionary Catholic cantons (the Sonderbund) and the founding of the new federal state.

From an early age Friedrich was recognized as being highly intelligent, but he was shy and introspective—perhaps in part as the result of a serious hearing impairment he had suffered from since boyhood. Despite this handicap he took a great interest in music, something he had in common with his father, a talented singer who gave public concerts. Not unexpectedly, Friedrich did very well at school. When he graduated from the Gymnasium, it was obvious that he should continue his studies at the university. Both his father and his maternal uncle Wilhelm His were leading professors on the medical faculty of the University of Basel, His having been appointed to the chair in anatomy and physiology in 1857. Wilhelm's influence was probably decisive as the young Friedrich Miescher determined to follow in the footsteps of his father and uncle, and that influence would continue throughout Friedrich's life.

It may be of interest to consider for a moment the general situation of contemporary medicine when Miescher took up his medical studies. The basic sciences of medicine—anatomy, pathology, physiology, and biochemistry—had made enormous progress. When scientifically based clinical research slowly emerged in the first decades of the nineteenth century, however, the practice of medicine was in a state of bankruptcy. Not only was the intellectual framework, based on the body humors that had ruled medical thinking for two millennia, in chaos and disintegration, but even venesection (bloodletting), the unshakable foundation of all therapy, was being questioned. (Paradoxically, it would enjoy its last heyday just before its final exit from the scene.)

At the great hospitals in Paris, leading physicians such as François Broussais, a former sergeant in Napoleon's armies, and his pupil Jean Baptiste Bouillaud let loose rivers of blood, particularly in their treatment of pneumonia. Leeches were used on an unprecedented scale; demand for the useful little animals was so strong that they had to be imported from all over Europe. Bouillaud, especially sanguinary, set an all-time record in bloodletting when he routinely drew at least two liters of blood from his unfortunate victims in a week's time. No wonder the mortality rose to such staggering heights at his clinic!

As Broussais and Bouillaud caused bloodshed of almost Napoleonic proportions, skeptics were questioning the merits of venesection—chief among them the French physician Pierre Louis. The whole therapeutic tradition, he had come to realize, especially all kinds of bloodletting, was not based on a critical evaluation of clinical results. It had simply been handed down since antiquity without ever having been seriously questioned by the medical community (with the exception of revolutionaries such as Paracelsus and van Helmont, of course). To put an end to this mindless tradition, he introduced *la méthode numérique;* that is, he relied on a simple statistical analysis of his therapeutic results. When he tried to implement these principles to evaluate the conventional treatment of pneumonia, he ran into difficulty. Because of the relentless activity of Broussais and his followers, there were no patients who had completely escaped bloodletting. He therefore had to be content with comparing patients who had been bled until they became unconscious (Bouillaud's favorite method) with those who had been bled only moderately. His results showed without any doubt that bloodletting, far from increasing a patient's chance of survival, made the prognosis considerably worse.

When Louis published his pioneering work in 1835, it was met with a storm of contempt and fury from the leading French clinicians. Not until 1849, when the Austrian physician Joseph Dietl confirmed and extended Louis' findings, were they finally accepted.

During the 1850s bloodletting disappeared from the medical scene as if through a trapdoor. In its fall it dragged with it the whole venerable edifice of Hippocratic medicine that had held the medical world enthralled for well over two thousand years. But its demise left a therapeutic vacuum, and physicians became the helpless victims of merciless caricaturists such as Honoré Daumier. Never before or since has the practice of medicine been held in such low esteem. In striking contrast, in the basic medical sciences rapid progress was being made in almost every field. It was clinical medicine that lagged in the most depressing way; consequently, many of the brightest students entering medicine stayed away from the intellectually less rewarding clinical work.

It is hardly surprising that during his medical studies, like his father and his idol Wilhelm His, the young Miescher should lean toward scientific research rather than the unpromising clinic, and no doubt his uncle encouraged him on this route. Miescher must have taken an interest in biochemistry, which was coming into its own as an independent science and not just a branch of physiology. In 1865, while he was still a medical student, Miescher went to Göttingen for the summer in order to work in the laboratory of the organic chemist Adolf Strecker. On his return to Basel he contracted typhoid fever and had to interrupt his studies for almost a year. Nevertheless, he got his M.D. in 1868 and considered what field in medicine he should enter.

Friedrich Miescher then did something that would surely boggle the mind of any young physician of our time. He wrote a long letter, or rather an extensive essay, to his father in which he set out his thoughts about his own future and carefully weighed the pros and cons of his various options. Even the effects of his hearing impairment were thoroughly considered, for this was a serious handicap indeed for a clinician. At a time when diseases of the lung, such as pneumonia and tuberculosis, claimed countless victims every year, percussion while listening through the stethoscope was the only way to examine the organs in the chest cavity. All these things Miescher carefully took into account, but it is obvious that what he really wanted was to take up a research career in biochemistry or physiology. However, the letter also showed a lack of confidence in his own ability, and he shrank from committing himself wholeheartedly to research.

In the end he came up with a compromise; he would first spend a semester in a biochemical laboratory and then devote two semesters to physiology. The compromise was that after this fairly extensive scientific training he intended to specialize in ophthalmology and, somewhat unexpectedly in light of his handicap, in otology. He considered that these specialties would leave him some spare time in which to pursue his scientific interests.

Like many other compromises, this one suffered from a serious and probably insoluble conflict of interests. We do not know the immediate reaction of Friedrich Miescher Sr. to his son's letter, but we do know that he forwarded it to Wilhelm His and asked for his opinion. Unlike Friedrich Miescher, His had no doubts whatsoever about his nephew's ability to become a good, perhaps even a great, scientist. The only critical comment His made was that Friedrich tended to be a bit sloppy in the laboratory and leave unwashed glassware and instruments lying about. In view of Miescher's utterly painstaking future research, he must have taken his uncle's criticism to heart. Otherwise His was full of praise for his nephew, whom he considered the best of all his students. At the same time His rejected Miescher's compromise out of hand. He could not see that it would make sense for Miescher to commit himself to ophthalmology or otology at this stage of his career. Rather, he urged him to go ahead with his plans for a period of training in biochemistry and physiology and not to worry about what would follow. Sufficient unto the day . . . .

His's letter to Miescher Sr. apparently settled the matter. Wilhelm His was only thirty-seven years old at the time, twenty years younger than his brother-in-law and only thirteen years older than his nephew. It appears that in many ways

Friedrich Miescher looked up to him as one does to an elder brother, and this gave His a unique position in the younger man's life. Furthermore, it was already evident that His was destined for a distinguished research career. His word carried great weight with his brother-in-law, who probably recognized His as his scientific superior. It was eventually decided that young Friedrich should go to a biochemical laboratory in Tübingen that was led by a rising new star, Felix Hoppe-Seyler.

Felix Hoppe was born in 1825, the tenth child of a minister. After losing both his parents at an early age, he was raised by his brother-in-law, a physician by the name of Seyler. When Felix Hoppe was thirty-nine years old, his guardian formally adopted him and Felix changed his name to Hoppe-Seyler. He completed his medical studies in Berlin, then practiced medicine for a few years before he accepted a position as head of the chemical department in Virchow's new Institute of Pathology in Berlin. Having worked mainly in analytical chemistry during his Berlin years, he moved to Tübingen in 1860 and quickly made a name for himself by his studies on hemoglobin and its binding of oxygen. His laboratory was located in the vaults of the old castle in Tübingen, overlooking the river Neckar. In later days Miescher would fondly remember the narrow rooms with their deep-set windows that reminded him of an alchemist's laboratory.

Wilhelm His had repeatedly stressed to his nephew the importance of understanding the chemistry and physiology of the cell. When Miescher arrived in Tübingen, he had probably already decided that he wanted to study the chemistry of cellular structures and in particular the cell nucleus. This emphasis on the cell and its structures was a general trend in contemporary medicine and originated with Rudolf Virchow, who had published his epochal work, *Die Cellularpathologie*, in 1858. Thus, Miescher's suggestions for his postdoctoral work were not particularly original and in fact dovetailed closely with Hoppe-Seyler's own research interests, which included cellular elements in the blood such as the leucocytes. It was agreed that Miescher should focus his efforts on the leucocytes. Then, as so often in biochemistry, a logistical problem arose: where to find leucocytes in sufficient quantity and of sufficient purity?

Laudable Pus

Surgery had always been haunted by a terrifying specter that seemed inseparably associated with all operations—surgical fever and its obstetric counterpart, puerperal fever. Infections of surgical wounds were such routine complications that doctors had come up with a theory of the good pus, *pus laudabile*, and believed the flow of pus to be a necessary means for the organism to rid itself of poisonous humors. The laudable pus was very different from the stinking fluid that oozed from a gangrenous wound and inevitably heralded the death of the patient (unless he could be saved by an amputation). Even under favorable conditions, in peacetime, and in well-equipped and adequately staffed hospitals, the mortality from major amputations was something like 50 percent; in time of war it was considerably higher.

Ignaz Semmelweis's heroic struggle against puerperal fever had been largely ignored by his surgeon colleagues, and streams of pus continued to flow in the wards long after Joseph Lister had introduced antiseptic methods in 1867, when he published his famous paper with the somewhat misleading title *On a New Method of Treating Compound Fractures.* Thus, it is not unusual that a supply of laudable pus was on hand at the surgical clinic of the university hospital in Tübingen when Miescher started his work on the cell nucleus.

Miescher had originally intended to study lymphocytes, but he soon realized that it would be impossible to get enough of these cells. Encouraged by Hoppe-Seyler, he instead focused on leucocytes, known to be the main cellular constituent of the laudable pus that could be obtained fresh every day from used bandages in the nearby hospital. The trick was to wash the cells from the bandages without damaging them. To this end Friedrich tried various salt solutions, but the cells swelled and gave rise to a highly viscous porridge that was impossible to handle. In hindsight it is easy to see that this occurred because he had extracted high-molecular-weight DNA from the damaged cells. Eventually he hit on a dilute solution of sodium sulfate as the best way to rinse well-preserved cells from the bandages. After filtration to get rid of tissue fibers, the cells were left to sediment to the bottom of the beaker; laboratory centrifuges were nonexistent in those days. When examined in the microscope, the leucocytes seemed intact and showed no signs of damage. He could demonstrate that what we call the cytoplasm was to a large extent made up of protein, which he could fractionate by precipitation with salt at different concentrations.

These results were encouraging but not sensational; that cells contained a variety of proteins was a well-established fact. The real enigma was the chemical nature of the nucleus. Miescher had previously observed that something could be extracted with weakly alkaline solutions from his porridge of swollen cells, something that precipitated when the extract was neutralized with acid. A similar phenomenon had been reported in the literature, and it had been suggested that the precipitate consisted of myosin, a protein that had already been found in muscle. The error goes back to the general tendency of the period to identify new proteins with others that were already known. However, Miescher soon satisfied himself that the precipitate was not myosin; instead, he felt sure that the unknown substance was derived from the nucleus.

His first task was to isolate undamaged nuclei free of cytoplasm. This had never been accomplished before and only after long hours of hard work did Miescher come up with reasonable quantities of nuclei in good condition. He tried several methods, but his final procedure was as follows. Miescher first treated the cells with warm alcohol to remove lipids, then digested away the proteins of the cytoplasm with the proteolytic enzyme pepsin. What he used was not a pure preparation of crystalline pepsin, as would be used today; nothing comparable was available. Instead, he extracted pig's stomach with dilute hydrochloric acid, which gave him an active but highly impure enzyme that he used for his digestions. The pepsin treatment solubilized the cytoplasm and left the cell nuclei behind as a grayish precipitate. The nuclei seemed entirely free of cytoplasm, and he exultantly concluded that with this method he could prepare any quantity of nuclei in a completely reproducible way. In his enthusiasm he exaggerated a bit about the quantity, but he was certainly in a position now to investigate the chemical composition of the nucleus.

He subjected his purified nuclei to the same alkaline extraction procedure he had previously used with whole cells, and on acidification he again obtained a precipitate. Obviously this material must have come from the nucleus, and he therefore named it nuclein. Using elementary analysis, one of the few methods available to characterize an unknown compound, Miescher found that his new substance contained 14 percent nitrogen, 3 percent phosphorus, and 2 percent sulfur. Its comparatively high phosphorus content and its resistance to digestion with pepsin suggested that the substance was not a protein. (At the same time, its appreciable content of sulfur immediately indicates to us today that his preparation did contain substantial amounts of protein.)

Even if Miescher was fully aware that he had discovered an entirely new group of substances that he thought might compare in importance with the proteins, he had no idea of the biological function of his nuclein. He suggested that its role was that of a storehouse for phosphorus, which could take up or dispense this important element as required by the cell.

Paradoxical as it may seem, Miescher to the end of his life rejected the idea that nuclein might have something to do with heredity. Later on we will consider this problem in more detail.

In the autumn of 1869 Miescher left Tübingen to join the laboratory of the famous physiologist Karl Ludwig in Leipzig. He had been working on a manuscript describing his discovery of nuclein, and in December of that year sent it to Hoppe-Seyler with a request that it be published in *Hoppe-Seyler; medicinisch-chemische Untersuchungen* (Medical-chemical investigations). He immediately ran into difficulties that set the tone for how the scientific community would receive his great discovery. Hoppe-Seyler found it hard to believe Miescher's results in spite of the fact that they had been obtained in his own laboratory. In those days fraud was not recognized as the bane of biomedical research, as it is now supposed to be, but Hoppe-Seyler may have had something like that at the back of his mind when he hesitated to publish the work. He even suggested that Miescher send his manuscript elsewhere, since it would be May until it could appear in the Hoppe-Seyler's journal. It seems that he was stalling for time in order to repeat Miescher's experiments and confirm his results. Having done so and satisfying himself as to the soundness of Miescher's work, Hoppe-Seyler finally accepted the paper. With all these delays, exacerbated by the general confusion that accompanied the war between France and Prussia in 1870–71, the paper did not appear until the spring of 1871, together with the confirmatory results of Hoppe-Seyler and a note explaining that Miescher's work had been completed in 1869 but its publication delayed by unforeseen circumstances.

Most modern scientists would have been infuriated by this procrastination, but Miescher seems to have taken it very calmly, to judge from the correspondence with his former mentor. He was not a pushy person, and perhaps his mind was occupied elsewhere now that he had joined Ludwig's laboratory. His new professor was one of the most celebrated physiologists in Germany, and young scientists from all over the world came to work with him. Miescher seems to have felt at home with this international crowd. He made several new friends among them, including Karl Ludwig himself, with whom he would stay in touch for the rest of his life. One has the feeling that the year Miescher spent with the dynamic and rather dominating Ludwig was personally a very satisfying period.

Ludwig kept his collaborators on a tight rein. They were not encouraged to go off on their own and pursue independent projects. The activities in the laboratory were strictly coordinated, with Ludwig very much in control. Miescher apparently did not object to this monolithic organization; in a letter to his parents he humorously described how he and his colleagues had only a vague idea of what went on in their leader's head. Occasionally they might, to their surprise, find a fine paper in the literature with their names on it, without always being able to say in what way they had contributed to it. Be that as it may, the stay in Ludwig's laboratory had a strong impact on Miescher's scientific thinking and no doubt profoundly influenced his future research career.

When the Salmon Spawned in the Rhine

In 1871 Miescher was back in Basel for his *Habilitation*, which would make him eligible as *Privatdozent*, an unsalaried position that was nevertheless a necessary step in his academic career. The process included presenting a lecture, and he chose as the subject for it not his discovery of nuclein, but work that had been done in Ludwig's laboratory. The next year his beloved uncle, Wilhelm His, left Basel to become professor of anatomy at the University of Leipzig, where he would remain for the rest of his life. From a personal standpoint this was undoubtedly a blow to his nephew, but at the same time it provided him with a great opportunity. His had been professor of both anatomy and physiology; when he left the university, two new chairs were created, one in each subject. On Miescher's return to Basel, his uncle had entrusted him with teaching duties in physiology and when the new chair in that subject became available, His used his influence to have Miescher appointed to it.

In Basel Miescher took up his old research on nuclein, but it became increasingly clear to him that to make real progress he needed a better and more abundant starting material than pus. Probably as the result of the interest that his uncle had taken as an embryologist in the development of the fertilized salmon ovum, Miescher hit on the idea of using salmon sperm as a source of nuclein. The sperm head was known to be the equivalent of the cell nucleus, and since it contained practically no cytoplasm it should be ideal for the preparation of nuclein.

Today, when we have managed to convert the Rhine into a sewer, it is difficult to imagine a time when the salmon went upriver to spawn. But in the 1870s salmon fishing in the Rhine was the basis of an important industry with its center in Basel; salmon sperm was easily available at practically no cost. Economic considerations were important to Miescher, who was hard pressed for space, equipment, and money to sustain his research. In a letter to a friend he admits that he often longed for the well-equipped and generously financed laboratory of Hoppe-Seyler. Even after he became professor of physiology his facilities were far from luxurious. The physiology department was housed in two rooms in an old building. As laboratory space for his own research, he had to be content with a corridor for his chemical analyses and a quarter-time technician to help him. No wonder he missed his alchemist's laboratory in the old castle of Tübingen!

Under these extremely primitive conditions, Miescher did the work that would establish the nature of nuclein, a new group of biological substances that would eventually, as Miescher had predicted, prove to be as important as the proteins. He found that after the lipids had been extracted, the isolated sperm head consisted almost entirely of a saltlike combination of nuclein and an organic base (a small basic protein, in today's terminology) that he called protamin. The word "nuclein" he now used to designate the equivalent of what is now called DNA, whereas he had originally used the term to indicate the mixture of DNA and protein prepared from cell nuclei. This ambiguous use of the word "nuclein" was to cause a certain amount of confusion in the literature until the German biochemist Richard Altmann introduced the term "nucleic acid" in 1889.

Miescher found that he could extract the protamine with dilute acid, which left behind an insoluble residue of nuclein. The protamine was purified and eventually crystallized as the hydrochloride, while the nuclein could be solubilized in dilute alkali and precipitated with alcohol after acidification. This new sperm nuclein had a phosphorus content of 9.6 percent, was free of sulfur, and contained no protein as judged by the color tests available at the time. In a famous letter to his uncle, Meischer describes in some detail the arduous preparation procedure, which had to be completed in one day: "Only greatest possible speed and low temperature leads to the goal. In order to prepare nuclein, I go to the laboratory at 5 o'clock in the morning and work in an unheated room. No solution may be kept for more than five minutes, and no precipitate left for more than an hour, before it is all preserved under absolute alcohol. The work often goes on until late at night."

No doubt His must have been pleased to hear about Miescher's progress, but at the same time he may have worried about the writer's health. In the letter to Miescher's father in which he discussed his nephew's future, His specifically warned against overwork and Miescher's tendency to neglect a young man's need for relaxation and an occasional day off from work and studies. Obviously, Miescher was dedicated to science to the point of being almost a fanatic.

In a speech at the hundredth anniversary of Miescher's birth, one of his former students, F. Suter, commented at some length on Miescher's dedication and said that he seemed to be driven by a demon. It was not so much a question of ambition and longing for fame and recognition as an almost religious conviction that research was paramount. To illustrate, Suter tells two anecdotes. When Miescher was short of glassware in the laboratory, he used his own Sèvres china instead, much to the despair of his unfortunate wife. Even worse, on their wedding day he did not show up in church at the appointed time. He had to be fetched from the laboratory, where he was working at the bench, having in his complete concentration forgotten all about his waiting bride. Stories like these are often told about great scientists and we may doubt their authenticity; but the fact that Suter related these tales certainly illustrates his own view of Miescher's personality.

The strong awareness of having a mission of overwhelming importance in life was coupled in Miescher with an equally strong feeling of inadequacy for the task he had been given. More than willing to exert himself to the utmost to achieve his goal, he still seemed perpetually insecure and uncertain of himself. He had been chosen, but was he really worthy of his formidable cause? His was aware of this lack of self-confidence and commented on it in his letter to Miescher Sr. With the faith in his own ability that his nephew so obviously lacked, His declared that he failed to see how anyone who worked seriously and energetically toward a specific goal could in the end fail to achieve something of importance. He went on to say that "Fritz," with his dedication and intellectual gifts, would certainly find both success and personal satisfaction in whatever task he undertook.

It is almost pathetic to observe these two men, so close to each other and so completely different. There is no mistaking the warm affection of Miescher for his admired uncle, and Wilhelm His, in many ways the stronger of the two, must also have been very fond of his moody and uncertain nephew. In 1874 His wrote a book titled *Our Bodily Forms and the Physiological Problem of Their Genesis: Letters to a Scientific Friend.* The friend was his nephew Friedrich Miescher. This rather strange way of publishing data had a famous precedent. Giovanni Battista Morgagni, the father of modern pathology, had published his lifework in 1761 at the advanced age of seventy-nine, in the form of a series of letters to a young colleague, whose identity has never been revealed. In any case, His's book goes a long way to show the fondness he felt for his nephew.

Miescher gave a preliminary account of his work with salmon sperm nuclein in 1872, when he read a paper at a meeting of the Society for Scientific Research in Basel. In 1874 he published a full account of the subject under the somewhat misleading title *The Spermatozoa of Some Vertebrates*, which appeared in the transactions of the society.

The Bewildered Pathfinder

The picture that emerges of Miescher as a man and as a scientist is puzzling and contradictory. He was a shy and intensely private person who does not seem to have made friends easily. Nor was he a great communicator. His students remembered him as an uninspiring, even awkward lecturer, in spite of the fact that he was a conscientious teacher and took great pains in preparing his lectures. His young collaborators in the laboratory respected him for his hard work and his dedication to science, but they do not seem to have been on intmate terms with him. On the other hand, in the extensive correspondence with his beloved uncle that continued uninterrupted for the whole of Miescher's adult life, he gave a detailed picture of his scientific activities and thinking; occasionally we also catch a glimpse of Friedrich Miescher, the man.

Certainly, Wilhelm His was the best friend his enigmatic nephew had—and perhaps the only one really close to him. At the same time, one cannot help wondering if Miescher's ability to reveal himself in these letters and let us see him as a three-dimensional human being has something to do with the fact that he expressed his thoughts and feelings in writing. One is reminded of the incredibly long letter that Miescher as a young man wrote to his father, setting out his detailed thoughts about his own future and carefully weighing the various options. He may have found it easier to communicate without restraint in writing rather than face to face. Perhaps his hearing impairment was a factor. In any case, we have every reason to be grateful that fate decreed that he and Wilhelm His should live in different countries and be obliged to keep in touch by letter.

One mystery surpasses all the others in Miescher's life. Why did he suddenly give up his research on nuclein, after publishing his 1874 article on the isolation of what we today call DNA? He had made a landmark discovery, and there can be no doubt that he himself realized that. Still, his publications had been met with indifference and in many cases with downright mistrust. Even his former mentor, Hoppe-Seyler, had been doubtful in the beginning. Miescher's results were questioned as early as 1874 by the German chemist Jakob Worm-Müller, and in 1878 the English chemist Charles Kingzett, having analyzed nuclein, brutally rejected it as "nothing but an impure albuminous substance." Citing Kingzett's report, Johann Thudichum concluded in 1881 that "nuclein was exploded, at least in this country." Similar opinions had been expressed by the French chemist Adolphe Wurtz a year earlier. Even in the late eighties and early nineties, several authors dismissed Miescher's nuclein as phosphorylated protein, a view that was reiterated by Edward Minchin in a review article in 1916.

These rejections occurred long before the time when the first priority for a successful scientist was to make the biggest possible splash in the media with an outstanding new discovery, but even if he were living in our own public-relations-conscious world, Miescher would not have been that kind of scientist. Yet the lack of appreciation and the harsh criticism from his scientific peers must have hurt him deeply. Were these his reasons for abandoning the research on nuclein and turning to entirely different projects? Surely, if his discovery of nuclein had been unanimously hailed as the great step forward in biomedicine that it eventually proved to be, he would not have changed fields! The cold reception of nuclein by the scientific community very likely depressed Miescher and caused him to lose interest in the subject, but other circumstances may also have contributed to his decision.

Biochemistry, or physiological chemistry as it was generally called then, was still regarded as a branch of physiology, something true chemists tended to look down on as less precise and scientific than, for instance, organic chemistry, which was enjoying enormous prestige, to say nothing of its growing economic importance. At the medical faculties, physiology was an old and well-established science; biochemistry was a newcomer and something of a poor relative of physiology. After all, the chair that Miescher had been appointed to in Basel was in physiology, and he may have felt that his first duty was to promote that subject at the university. Since His, his predecessor, had been primarily interested in morphology and embryology, there is every reason to believe that physiological research was in need of a boost when Miescher succeeded his uncle. The conscientious Miescher may well have felt that his first obligation was to create a strong school of physiology in Basel to carry on the traditions he had assimilated in Ludwig's laboratory.

Whatever prompted Miescher to make this decision, he

dropped nuclein completely and did not return to it until a few years before his untimely death. Instead, he studied the physiology of the spawning salmon, particularly the relation between the tremendous increase of its sexual organs and the simultaneous degeneration of its trunk muscles. Later on he became interested in the physiology of respiration. After the department moved into a new building with vastly improved research facilities, respiration became his main line of research. Miescher also became involved in nutritional problems and was asked by the city government to report on nutritional conditions in the local prison. To his dismay, he came to be regarded as something of an expert on nutrition; in a letter to His he complained that he had become "the watchdog of the stomachs of three million fellow countrymen."

When Miescher was thirty-four years old he married Mary Rüsch, who was twelve years younger than he. If we are to believe her obituary in 1946, she was extremely shy and found it difficult to put her feelings into words. Miescher apparently chose a wife in his own image. We do not know much about his married life, but he seems to have been a devoted husband and father, albeit somewhat distracted by his preoccupation with science. His wife survived him and their three children by many years, but her life after Miescher's death was filled with misfortunes and unhappiness. Their elder daughter, Frieda, died when she was only twenty-two, and their son, Fritz, succumbed to complications following an operation at the age of thirty-three, while doing postdoctoral work in Berlin after having completed his medical studies. The worst blow was the fate of the younger daughter, Maria, who shortly after her brother's death became mentally ill. Eventually she had to be committed to an institution when her mother could no longer care for her.

Friedrich Miescher's own health was never robust, and in 1885 he suffered from a pleurisy that, in view of his later illness, probably was tubercular in nature, and in 1894 tuberculosis of the lungs was diagnosed. Friedrich was sent to a sanatorium in Davos, where he remained until his death a year later, on August 26, 1895. His illness progressed rapidly and Miescher, who made heroic efforts to stay in close contact with his collaborators and direct the research of his group, finally had to resign his chair. In a letter Miescher described how the verdict of his doctor that he would never be able to resume his work was a staggering blow. That blow may have been softened to some extent by a kind letter from the head of the local government (*Regierungspräsident*), Isaak Iselin, who thanked him for his services to the university.

Characteristically, Miescher in his reply professed to have done nothing more than could be expected of any citizen. He ended his letter by saying that he assumed the government had merely wanted to give comfort and pleasure to a seriously ill citizen by its words of praise. This may seem like false modesty, but sadly enough Miescher probably meant what he said. An often-cited letter to an old friend from the happy Leipzig days described how every night he went to bed with the feelings of a schoolboy who had not done his homework. As he lay dying in Davos, he may have looked back on his life and felt he had not accomplished what he set out to do. The self-doubt and lack of confidence that his uncle had long ago detected in Miescher asserted itself even at the end.

It is ironic and sad that in the eulogies after Miescher's death less weight was given to his discovery of nuclein than to his achievements in physiology and nutrition. At the decisive crossroad in his scientific career, he had chosen the wrong path; in bewilderment, he lost sight of what should have been his main goal. As we shall see, he went astray also in that he refused to believe nuclein could have anything to do with heredity.

Nevertheless, the credit for having discovered DNA belongs to Friedrich Miescher, the least pretentious of scientific heroes. One man at least realized it: Karl Ludwig, his former professor in Leipzig. In a moving letter to Miescher just before his death, Ludwig paid tribute to the lasting importance of Miescher's discovery of nuclein. The letter ended with the words, "When in coming centuries men work on the cell, your name will be gratefully remembered as the pioneer in this field."

Miescher's discovery of DNA had been greeted with suspicion and indifference by his biochemical colleagues. The cell biologists, however, who represented a new and vigorous branch of biomedicine, took an entirely different view. Unlike Miescher himself, they recognized the true function of his nuclein. This insight was based on their study of cell structures demonstrable by the use of new staining techniques.

Paracelsus 1493–1541) Courtesy of Bonnierförlagen, Stockholm, Sweden.

Johann Baptista van Helmont (1577–1644) Courtesy of Bonnierförlagen, Stockholm, Sweden.

John Mayow (c. 1640–1679)
Courtesy of the Royal Society,
London.

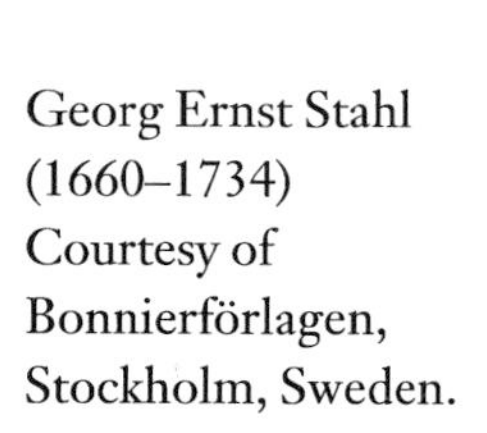

Georg Ernst Stahl
(1660–1734)
Courtesy of
Bonnierförlagen,
Stockholm, Sweden.

Antoine-Laurent Lavoisier (with his wife) (1743–1794)
Painting by David. The Metropolitan Museum, New York.

New York.
Justus von Liebig
(1803–1873)
Courtesy of the Royal Swedish Academy of Sciences, Stockholm.

Friedrich Wilhelm Schelling
(1775–1854)
Courtesy of Bonnierförlagen, Stockholm, Sweden.

Louis Pasteur
(1822–1895)
Courtesy of the Royal
Swedish Academy of
Sciences, Stockholm.

Jöns Jakob Berzelius
(1779–1848)
Courtesy of the Royal
Swedish Academy of
Sciences, Stockholm.

Richard Willstätter (1872–1942)
Courtesy of the Nobel Foundation, Stockholm, Sweden.

Friedrich Miescher (1844–1895)
Courtesy of the Friedrich Miescher-Institut, Basel, Switzerland.

Wilhelm His (1831–1904)
Courtesy of the Friedrich Miescher-Institut, Basel, Switzerland.

A Guild of Dyers

The second half of the nineteenth century and the first decades of the twentieth were the golden age of German organic chemistry. Such luminaries as Friedrich von Kekulé, Adolf von Baeyer, Emil Fischer, and Richard Willstätter led the field to scientific eminence and a dominating role in the German chemical industry. The synthesis and industrial production of new dyes was of major economic importance for the country and undoubtedly contributed to the increasing political power of the German Empire. Companies like I. G. Farben became world leaders in synthetic dyes, and the growth of the industry had side benefits also for the biological sciences. It was rapidly discovered that the new dyes could be used not only on textile fibers to create much more colorful women's fashions, but also on cellular structures in both the nucleus and the cytoplasm.

The staining properties of different industrial dyes were systematically investigated in the 1870s by Paul Ehrlich, better known for his later contributions to immunology and his introduction of salvarsan for the treatment of syphilis. It was found that the nucleus contained structures that were readily stained by basic dyes, and Ehrlich called these structures basophilic. In contrast, the cytoplasm was found to be preferentially stained by dyes with an acidic character. In 1882 the German histologist Walther Flemming published his seminal book, *Cell Substance, Nucleus, and Cell Division*, in which he summarized what the new dyeing techniques had proved concerning the cell nucleus. He introduced the term "chromatin" to denote the basophilic structures in the nucleus and suggested that they were identical with Miescher's nuclein—or at least contained nuclein as part of their structure—and that this was the reason they stained with basic dyes.

At this time the function of the nucleus still remained something of an enigma, even if the German biologist Oscar Hertwig had observed in 1875 that, on fertilization of the sea-urchin egg, the sperm nucleus united with the egg nucleus. Subsequent cleavage of this united structure gave rise to the nuclei of the following cell generation in the development of the embryo. This basic phenomenon was later observed both in other animals and in plants. Flemming investigated the structural transformation of the nucleus during cell division and introduced the term "mitosis" to denote this process.

In 1883 Edouard van Beneden published his studies on the fertilization of eggs from the worm *Ascaris megalocephala*. He found that when the sperm and egg nuclei united, the chromatin in the fusing nuclei was organized into a series of rod-like structures of the same shape and number. This observation was soon extended to other animals as well as plants, and the morphological structures discovered by van Beneden were later termed "chromosomes." The characteristic behavior of the chromosomes during cell division—their longitudinal cleavage and the distribution of the replicated chromosomes among the daughter cells—led Wilhelm Roux to postulate that they were the bearers of heredity. This proposal was seconded by the eminent biologist August Weismann, who had been the first to realize that in a multicellular organism the germ cells are the exclusive carriers of genetic information from one generation to the next.

It was inevitable that chromatin, and more specifically the nuclein it contained, should be proposed as the molecular basis of heredity. In 1885 Oscar Hertwig had suggested that "nuclein is the substance responsible not only for fertilization but also for the transmission of hereditary characteristics." Similar views had been expressed by others. The well-known American biologist Edmund B. Wilson summed up the prevailing views of contemporary biology when he wrote in 1896 that "the chromatin is to be regarded as the physical basis of inheritance. Now chromatin is known to be closely similar to, if not identical with, a substance known as nuclein . . . , which analysis shows to be a tolerably definite chemical compound of nucleic acid (a complex organic acid rich in phosphorus) and albumin. And thus we reach the remarkable conclusion that inheritance may, perhaps, be effected by the physical transmission of a particular compound from parent to offspring."

Where did Meischer stand in this animated discussion of the biological function of nuclein? Surely he must have sided with those who identified nuclein as the carrier of genetic information! Its leading role ought to have compensated him for the initial scorn and mistrust that the great discovery of his youth had generated. This would seem an obvious conclusion, but in reality Miescher rejected the whole theory of nuclein as the genetic material. What were his views, then, and how can we understand this apparent paradox?

In his publication on sperm nuclein in 1874, Miescher discussed a possible role for it in heredity, only to dismiss this as improbable. He noted that sperm cells had been suggested as carriers of fertilizing chemical substances and admitted that if one accepted this idea, "one must undoubtedly think above all on nuclein." However, he went on to reject the concept of specific fertilization substances. He hinted instead at what he called a higher explanation and pictured "an apparatus that brings about or transforms some kind of motion."

To understand Miescher's surprising, not entirely lucid views, consider the theories of fertilization that were widely held at the middle of the nineteenth century. They were, in fact, closely related to the ideas about fermentation as caused by molecular vibrations transmitted from one substrate molecule to another. Central to these theories of fertilization, as expounded for instance by Theodor Bischoff in 1847, was the belief that this process was best described in terms of "contact" and "catalytic force" and did not involve the transfer of chemical substances from the sperm to the egg. It is probably significant in Miescher's case that his beloved uncle, Wilhelm His, in his book of 1874 had taken the stand that nothing substantial was transferred in fertilization. His visualized the contact between sperm and egg as transmitting "the excitation to form-developing growth." As a consequence of this somewhat opaque reasoning, His downplayed the role of the nucleus in processes such as fertilization and development.

To the end of his life Miescher continued to reject the idea that nuclein was the genetic material. He was not impressed by the evidence presented by cell biologists using the new dyeing techniques for the exciting morphological transformations of chromatin during fertilization and cell division. In 1890 he wrote to his uncle: "I must again defend my skin against the guild of dyers who insist that there is nothing but chromatin (nuclein), even if the presence of various spatially separated substances in the sperm head can be demonstrated easily."

Yet Miescher seems to have modified his previous view that there were no specific fertilization substances. In another letter to His in 1892, he hypothesized that an iron-containing protein, "karyogen," which he claimed to have isolated from the sperm head, was the important factor in fertilization and was also responsible for the basophilic coloring characteristics of chromatin; nuclein was just an envelope for the karyogen. Miescher's views about karyogen must be understood as implying that this protein was, in fact, the specific fertilization substance whose existence he had previously denied. However, the presence of this hypothetical iron-protein in the sperm head has never been confirmed by other investigators and must be regarded as an experimental artifact. In fairness to Miescher, it should be emphasized that he never published this finding.

In letters to His in 1892–93, Miescher proposed a rather fanciful theory about the stereochemistry of proteins as the molecular basis for heredity. He again rejected what he called "the speculations of Weismann and others" as being "afflicted with half-chemical concepts that are partly unclear and partly correspond to an obsolete state of chemistry." He pondered the vast number of isomeric forms of an assumed protein, which he inferred from the number of asymmetric carbon atoms in its molecule, and he considered the possible role of these isomers as the structural basis for genetic information. In retrospect, it was not Weismann and his guild of dyers who were the victims of "half-chemical concepts," but Miescher himself.

Right after the turn of the century, this theory of chromatin/nuclein as the genetic material, which had dominated biology at the end of the nineteenth century and irritated Friedrich Miescher, started to lose favor in the scientific community. Why was this essentially correct theory suddenly regarded as obsolete, and what made it fade into obscurity for almost half a century before it was resurrected? As is so often the case in science, many factors were involved.

One element was undoubtedly the apparent lack of temporal continuity of chromatin in the nucleus. It sometimes seemed to disappear, as in the case of certain chromosomes that had apparently lost their ability to be stained by basic dyes. This failure was taken as an indication that they no longer contained any nuclein. Outstanding cytologists such as Eduard Strasburger therefore believed that chromatin could not be the hereditary substance, since it occasionally seemed to be absent from the chromosome, and the amount of it in the nucleus varied considerably depending on the stage of development (Strasburger, 1909). In reality, DNA was not absent from these structures; it had only become inaccessible to the dyes used to demonstrate its presence.

Another argument against chromatin came from the picture of DNA as a comparatively small molecule with a monotonous structure. This image emerged at the beginning of the twentieth century as the result of structural studies to be considered in the next chapter. Such a molecule appeared rather unattractive as the structural basis of genetic information, whereas proteins with their unbelievably high molecular weights and complicated structures seemed altogether more likely candidates. Consequently, in the third edition of his book (which appeared in 1925, twenty-nine years after the first edition), Edmund Wilson completely rejected the views on nuclein and heredity he had expressed in the first edition and stated, "These facts offer conclusive proof that the individuality and genetic continuity of chromosomes do not depend on the persistence of chromatin in the older sense." He implied that the loss of stainability "seems to indicate a progressive accumulation of protein components and a giving up, or even a complete loss, of nuclein." Friedrich Miescher would have liked that statement, had he lived long enough to read it.

Almost fifty years after Miescher's death, the truth was finally established. He was proved wrong in his views about nuclein and heredity, while at the same time the fundamental importance of his remarkable discovery was made clear, once and for all.

[four] The Building Blocks

the old city of Rostock on an estuary of the river Warnow was founded in the twelfth century, when the Obotrite princes who ruled Mecklenburg were subjugated by Henry the Lion, Duke of Saxony, and made to embrace Christianity. In the fourteenth century Rostock became a member of the powerful Hanseatic League and was one of its leading cities. Although its zenith economically and politically was in the fifteenth and sixteenth centuries, it retained its position as a semi-independent city state until the end of the nineteenth century.

The Inheritor

It was here that Albrecht Kossel was born in 1853, the son of a well-to-do merchant and bank executive. He grew up in a large and apparently happy family: Albrecht had seven sisters and one brother, Hermann, who also became a physician and a scientist.

Even as a schoolboy Albrecht Kossel showed an interest in the natural sciences, botany and chemistry among others. Upon graduation from the Gymnasium in 1872 he entered medical school at the University of Strasbourg. Both Adolf von Baeyer and

Felix Hoppe-Seyler were active at the university, and Baeyer had just been joined by the graduate student Emil Fischer. This inspiring atmosphere, and in particular the contact with Hoppe-Seyler, must have influenced Kossel and strengthened his previous inclination toward chemistry. He returned to his native Rostock to complete his medical studies and graduated from medical school at its venerable university, which had been founded in 1419. It would seem that he had already decided on a career in biochemistry; in 1877 he returned to Strasbourg and joined his former teacher Hoppe-Seyler in what was then the leading biochemical laboratory in Germany.

In the same issue of *Hoppe-Seyler, medicinisch-chemische Untersuchungen* that contained Miescher's pioneering paper, the editor had also published his own discovery of a nuclein from yeast. This claim was challenged in 1878 by the well-known biologist Carl von Nägeli and his collaborator Oscar Loew, who dismissed Hoppe-Seyler's yeast nuclein as nothing but albumin contaminated by inorganic phosphates. Kossel picked up the gauntlet on behalf of his master and published a rejoinder in which he confirmed the presence in yeast cells of a substance rich in phosphorus and with the same solubility characteristics as nuclein. This was his first of many contributions to the nucleic acid field.

He could demonstrate that nuclein from several different sources seemed to contain hypoxantine and xanthine, substances known to be related to uric acid, whereas a phosphorus-rich material from egg yolk, which Miescher had mistak-

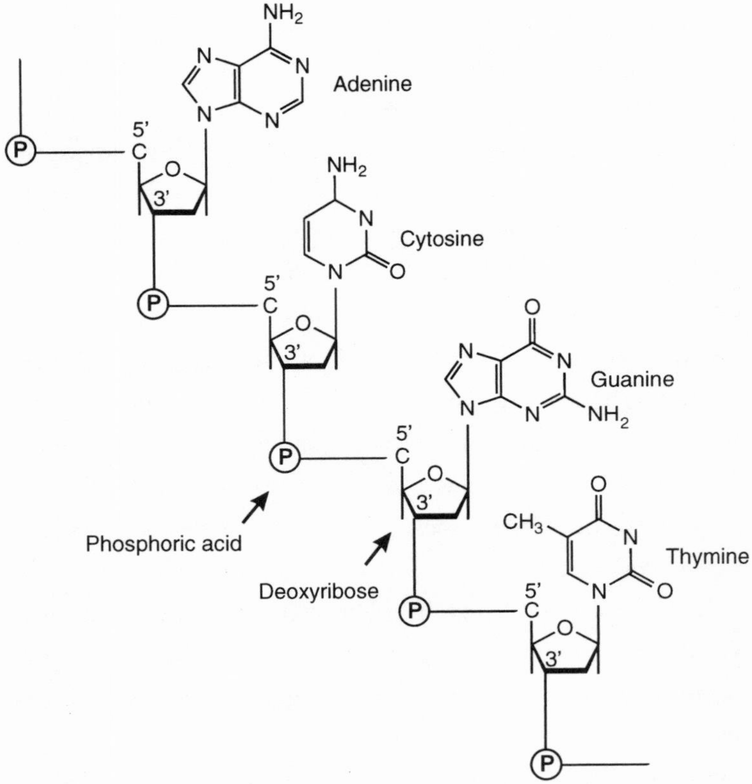

Figure 1 Building blocks of a nucleic acid: a simplified version of how a chain of nucleotides is built from its constituents, as exemplified by a DNA chain. The same general features are found in RNA. Each nucleotide contains a cyclic molecule (base) with carbon and nitrogen in the ring (a heterocyclic ring). The base is either a pyrimidine (cytosine and thymine in DNA; or, in the case of RNA, cytosine and uracil) or a purine (adenine and guanine). In DNA the nucleotides contain the sugar deoxyribose and are called deoxynucleotides, while in RNA the sugar is ribose. Both types of nucleic acids contain phosphoric acid, represented by the symbol P. In both DNA and RNA the nucleotides are linked to each other by phosphoric acid, which acts as a bridge between

enly believed to be nuclein, did not give rise to such substances on hydrolysis. This egg yolk "nuclein" would later be identified as a phosphorus-containing polypeptide, phosvitin. It had previously been claimed that hypoxanthine was a constituent of proteins, but Kossel could show that carefully purified preparations of protein did not yield hypoxanthine. He went on to demonstrate that so-called xanthine bodies, including adenine, guanine, and hypoxanthine, could be isolated following acid hydrolysis of nuclein. When after the turn of the century it was realized that hypoxanthine was a degradation product of adenine, resulting from the loss of an amino group, and that xanthine was related to guanine in the same way, it became apparent that adenine and guanine were constituents of all nucleins (Figure 1).

In 1889 Richard Altmann developed a method for the preparation of nuclein essentially free of protein and introduced the term "nucleic acids." Such "protein-free" nucleic acid from thymus was analyzed by Kossel and his collaborator Albert Neumann after hydrolysis with strong acid, and in 1893 they obtained a new product that they called thymine. It would appear from Miescher's posthumously published laboratory notes that he had isolated the same compound. Thymine obviously belonged to a different chemical group than the "xanthine bodies," or "purines" as they were now

---

adjacent nucleotides, as shown in the figure for a chain of deoxynucleotides. The bridge goes from the fifth (5′) carbon atom in the sugar moiety of one nucleotide to the third (3′) sugar carbon of the next nucleotide residue.

called by Fischer (who was in the process of elucidating their structure). These new compounds were called pyrimidines and in 1901 Kossel's associate Hermann Steudel determined the structure of the first pyrimidine—thymine. In 1894 Kossel and Neumann found yet another pyrimidine, which they named cytosine, in the hydrolysate of thymus nucleic acid.

Like adenine and guanine, cytosine was present in nucleic acids from both animal sources, such as thymus, and yeast. However, in 1900 Kossel's associate Alberto Ascoli demonstrated the presence of the pyrimidine uracil in hydrolysates of yeast nucleic acid, and a few years later another of Kossel's previous collaborators, Phoebus Levene, established that yeast nucleic acid contained uracil and cytosine but was devoid of thymine. It would seem that nucleic acids from thymus and yeast represented different classes of these compounds, a conclusion supported by results obtained independently by Kossel and by the Swedish biochemist Olof Hammarsten in the 1890s. They could demonstrate the presence in yeast nucleic acid of a carbohydrate that Kossel in 1893 identified as a five-carbon sugar, a pentose, based on its conversion to furfural by heating with strong acid (a reaction characteristic of pentoses). The carbohydrate component of thymus nucleic acid, on the other hand, under these conditions yielded not furfural but another compound, levulinic acid. It would, in fact, be a long time before Levene could demonstrate that the sugar moiety in thymus nucleic acid was also a pentose, deoxyribose, which differed from the sugar in yeast nucleic acid, ribose, by having one hydroxyl exchanged for a hydrogen atom (Figure 1). The nucleic acid that contains deoxyribose we now call deoxyribonucleic acid or DNA, while the ribose-containing variety is referred to as ribonucleic acid or RNA. We have also realized that all cells contain both kinds of nucleic acids, a fact that was not apparent at the time, when the names "thymus nucleic acid" and "yeast nucleic acid" were widely used to differentiate the two.

Kossel stayed with Hoppe-Seyler until 1883 and always spoke of his years in Strasbourg as a very happy period. However, when he was offered a post in Berlin as director of the chemistry division at the institute of the famous physiologist Emil DuBois-Reymond, it was an offer that he could not refuse. At the same time, Kossel seems to have had a dislike of everything Prussian, perhaps a legacy of growing up in the free Hanseatic city of Rostock, where the domineering Prussian ways were not exactly popular. Nevertheless, the twelve years he spent in Berlin were certainly productive scientifically, and it was there that he discovered that Miescher's protamine had a counterpart in ordinary cell nuclei. He called it histone and demonstrated that it existed in a salt-like combination with nuclein. The histone would later prove to be made up of several small, basic proteins. Kossel came to devote much attention to the basic amino acids that are found in substantial amounts in both protamines and histones.

The concept of building blocks was dear to Albrecht Kossel, and his enthusiasm for this idea sometimes carried him a bit too far. Nevertheless, his elucidation of the chemical nature of some of the building blocks that make up nucleic acids and chromatine has secured immortality for this exceedingly modest and almost shy man. He can be said to have inherited Miescher's great discovery, and it is amusing to see to what extent the personalities of the two men resembled each other. Not only were they similar in character, but they even had the same scientific mentor.

In 1886 Kossel married Luise Holtzmann, the daughter of a distinguished philologist, a charming and intelligent woman whose interest in literature and the arts was in sharp contrast to Kossel's relative indifference to all human activities other than science. In spite of their very different personalities—she was extroverted and an excellent hostess with a gift for witty conversation, while he was rather shy and not very talkative—theirs was obviously a happy marriage. Kossel had become an increasingly famous scientist and when his former mentor, Hoppe-Seyler, died in 1895, he took over as editor of what would henceforth be known as *Hoppe-Seyler's Zeitschrift für physiologische Chemie.* He continued in this capacity for the rest of his life. The Kossels had recently moved to Marburg, an attractive little town in Hesse. There he became professor of physiology at the university, which could boast such luminaries as Emil von Behring on its medical faculty.

Kossel had the ability to attract talented students and collaborators to his laboratory, many of them from abroad. (P. A. Levene would become the most famous.) His students at the medical school seem to have appreciated Kossel as a lecturer in spite of the fact that he was far from a natural orator. He compensated by preparing his lectures very carefully in the same way Miescher had done.

Kossel had indeed become the inheritor of Miescher, but it is doubtful that the two men had much contact with each other. Despite the obvious similarities in personality, Kossel seems to have lacked the self-critical trait that was so striking in Miescher. Although modest and unassuming, Kossel never wavered in the pursuit of his scientific goals as Miescher had. His former students spoke with admiration of his equanimity and the fact that they never saw him annoyed or irritated. Anyone who has been experimentally active and knows the aggravations that arise concedes this as a remarkable testimony to Kossel's cheerful nature and mental stability.

Another notable difference in the lives of these two leading figures is the way their discoveries were received by the scientific community. Miescher, who had after all opened up the whole field of nucleic acid research, was never recognized as a pioneer during his lifetime, whereas Kossel was showered with honors. In 1901 he was called to the prestigious chair of physiology at Heidelberg. Rumor has it that he accepted the position because he wanted to get out of Prussia. He received several honorary doctorates and in 1907 was made a *Geheimrat*, a title that brought with it certain valuable benefits. Three years later came the highest accolade, when he was awarded the Nobel Prize in Physiology or Medicine for his work on the chemistry of the cell nucleus. He continued to attract gifted young scientists from abroad, and his international fame and worldwide scientific contacts were a source of much satisfaction to him.

His political views seem to have been liberal, and the wave of chauvinism that followed the outbreak of World War I was repugnant and shocking to him. The year before, he had suffered a grievous loss when his wife died after a short illness. He felt isolated and depressed without her companionship and at the same time was deprived of the international contacts that were so important to him. He worried about the increasing intellectual isolation of German science and stubbornly refused to participate in "patriotic" manifestations, as other German professors did. Nor would he cooperate when summoned by the authorities to make optimistic pronouncements about the state of nutrition of the population, when the Allied blockade had made food scarce in Germany.

After the war it was a joy to him to renew his international contacts and participate in scientific meetings abroad. At the Eleventh Physiological Congress in Edinburgh in 1923, he was greeted as one of the leading delegates and given an honorary degree at the university. His almost childish delight at this recognition of a German scientist by the former enemy is quite touching.

There is something very attractive about this simple and honorable man. When he died peacefully at the age of seventy-four, one of his previous collaborators summed up his life by citing the words of the German neoclassicist Johann Winkelmann: *Edle Einfalt und stille Grösse* (Noble simplicity and quiet greatness). Not a bad epitaph for a scientist.

The Great Chemist

In 1918, a year before his death, Emil Fischer wrote a short autobiographic sketch that was published posthumously under the title *Aus meinem Leben*, with the following short note on the flyleaf: *Geschrieben in dem Unglücksjahre 1918* (written in the unhappy year 1918). He had every reason to feel depressed when he wrote this first draft of what he intended to be his memoirs, during a trip to Locarno and Karlsbad to regain his health. A widower for twenty-three years, he had lost two of his three sons under tragic circumstances, his health was in ruins, and his country was obviously losing the war. Taking all this into account, one might think his memoirs would read like something out of the Book of Job, but that is far from the case. Furthermore, given the commanding, even overwhelming, stature of Emil Fischer in the history of organic chemistry as well as biochemistry, one might expect him to have profound things to say about science, its role in society, and its relation to philosophy and ethics. However, one looks in vain for anything like that in this little book.

Fischer's memoirs are far from penetrating and sophisticated. They are naive and spontaneous, as if he had just written down whatever came into his head. He prattles on happily about the world of his childhood, his parents, his sisters (he was the only boy of eight children, and the youngest of them all), his uncles and aunts, and his innumerable cousins. At first one is almost shocked by the shallowness. Where is the keen mind that left its indelible mark on the chemistry of a whole generation? But in the end, one is captivated by the very naivete of these recollections.

Emil Fischer came from a Protestant family that had lived since the end of the seventeenth century in the small village of Flamersheim, in the foothills of the Eifel Mountains on the left bank of the Rhine. The family had come close to expiring on the male side, when Fischer's grandfather refused to marry, preferring to live with his tyrannical sister as his housekeeper. He changed his mind when she one day refused to give him the key to the wine cellar. In his understandable indignation at this high-handed treatment of a brother, at the mature age of forty-nine he resolved to marry. He did so and in fact had five children by his wife before she died in childbirth. When he followed her into the grave a couple of years later, Fischer's father was left an orphan and came into the care of an aunt in Mülheim on the Rhine.

Laurenz Fischer plays an important part in the autobiography of his famous son, and he is indeed something of a benevolent father figure. He was a big, burly fellow with robust health, who lived to be ninety-five years old. At school he was not very successful, dropping out at the age of fourteen, but he had lots of common sense and was shrewd enough to become a prosperous businessman on a modest scale. He loved hunting and other outdoor activities. A kindhearted, good-natured man who could never bring himself to punish his children, he seems to have left that to his wife, Julie, who was of a somewhat sterner disposition. Laurenz was a man of liberal views, an atheist and a hedonist who enjoyed life to the full to the last day of his life. In fact, Emil Fischer recalls that on the day of his father's death, Laurenz emptied a last tankard of his favorite beer, wiped his moustache, and said, "It's a good thing the common man can have such a fine beverage for so little money!" An hour later he was dead.

Julie Fischer was in many ways the opposite of her husband, but their marriage seems to have been felicitous. She was a much more serious person, with an intellectual turn of mind, and deeply religious. Julie was a staunch conservative and admirer of Otto von Bismarck. During his conflict with the Prussian Parliament in the sixties she was outspoken in her support of Bismarck, reproaching the rest of the family, who in her opinion were too stupid to understand and appreciate the great man. This made her husband, no admirer of Prussian Junkers or the Iron Chancellor, address her good-humouredly as "Frau Bismarck." There can be no doubt that Emil Fischer revered and respected his mother, that as a boy he probably stood in awe of her, but one has the impression from his memoirs that most of his affection was reserved for his jolly, easygoing father.

Emil Fischer was born in 1852 at Euskirchen, a small town to the north of the Eifel Mountains. His father had moved his business there, and they lived in a big house next to Laurenz' brother, who was also his business partner. Emil's childhood seems to have been a happy one. If his many elder sisters tended to be overprotective and coddle him too much, he could always escape to his uncle's house and have a refreshing fight with his male cousins. Unlike his father, Emil was something of a star pupil at school; perhaps he inherited this intellectual bent from his mother. In 1869 he graduated from the Gymnasium in Bonn with good grades, but in his memoirs he takes a critical view of the teaching there, with its emphasis on the mindless drilling of grammar and its neglect of both natural science and such humanities as literature and cultural history.

At age seventeen Emil had to choose a profession, and his father persuaded him to try a business career. He was apprenticed to his brother-in-law Max Friedrich, who had a timber business, and employed there as an errand boy—a position that did not appeal to him at all. He took to performing chemical experiments in a small laboratory that he had established for himself in an empty room at the factory, much to the annoyance of the foreman, who was understandably afraid that the amateur chemist would cause a fire. His brother-in-law was far from satisfied with him and declared to the rest of the family that he had never had such a useless apprentice, finishing with the harsh verdict, "That boy will never amount to anything." Father Laurenz agreed with his son-in-law and concluded, "The boy is too stupid to become a merchant; he should study!"

Having thus tried his hand at business, Emil Fischer went to the University of Bonn to study natural science, with an emphasis on chemistry. His own inclination was toward physics, but his father considered that subject altogether too theoretical to offer an economically secure future. Emil did not begin his university studies until 1871, however, having been delayed by a persistent gastritis, which he himself put down to heavy smoking and an excessive fondness for beer. He would, in fact, suffer similar complaints for the rest of his life.

At the university he came under the influence of the renowned chemist August von Kekulé, whose lectures he found highly stimulating. The rest of the chemistry courses bored him to tears, particularly quantitative inorganic analysis, which involved exceedingly tedious washings of various precipitates. After forty-six years Fischer still remembered having washed a certain precipitate of aluminium hydroxide for more than a week (there were no suction pumps in the chemistry department) without getting it free of the mother-liquor. In his severe frustration he was ready to give up chemistry, but was persuaded by his cousin Ernst to transfer to another university instead. Organic chemistry has reason to be grateful to Ernst for intervening at this critical stage of Emil Fischer's career.

Together with another of his many cousins, Otto Fischer, who would also become an outstanding chemist, he went to Strasbourg, which since the Franco-Prussian War was a German city. At the university there he studied with Adolf von Baeyer, one of the leading chemists in Germany. The contact with von Baeyer rekindled his fading enthusiasm for chemistry and had a decisive influence on his development as a scientist. He seems to have really enjoyed life in Strasbourg. His gastritis disappeared completely, for which he credited the French cuisine and the excellent wines. He worked hard all the same, graduating in 1874. When von Baeyer the next year took up a new position in Munich, Fischer accompanied him there. He recalls in his memoirs that his mother was very concerned because typhoid fever was rife in the city, and there had been an outbreak of cholera two years earlier. In fact, Fischer and his colleagues from Strasbourg sought the advice of a fellow scientist at the department of hygiene, who had a map of Munich on which he had plotted all the recent typhoid cases. Guided by this map, they selected suitable quarters in the least-stricken areas and were soon hard at work in the laboratory.

The most important result of Fischer's efforts in Munich was the discovery of a new class of highly reactive compounds, the hydrazines, which would prove to be useful reagents in his later work on carbohydrates. He was something of a favorite of von Baeyer, who recognized him as a rising star and did much to encourage and support him. Life in Munich obviously agreed with Fischer; he spoke longingly of the numerous *Bierkeller* and their relaxed atmosphere, where both beer and conversation flowed freely. In 1878 followed his *Habilitation* and his appointment as *Privatdozent*, an important step up the ladder to higher academic positions.

A year later von Baeyer saw to it that Fischer was promoted to associate professor and put in charge of analytical chemistry. That had not been his favorite subject during his chemical studies in Bonn, but now he was delighted to accept the new position, which considerably increased his resources at the university. He was only twenty-seven years old and his scientific career was already off to a flourishing start. Although his research was going very well, problems were waiting around the corner.

Ventilation in chemical laboratories was almost nonexistent in those days, and poisoning by chemical reagents was only too common. In the summer of 1881 Fischer had an attack of acute mercury poisoning as a consequence of having studied the effect of mercuric oxide on hydrazines. He seems to have recovered after a few months, and in 1882 was appointed full professor of chemistry at the University of Erlangen. There he had another bout of poisoning, this time caused by prolonged exposure to phosphochlorides—at least that is his own explanation of a persistent bronchitis from which he suffered for almost a year (one suspects that his heavy smoking did not help, either). Ten years later he would become the victim of chronic poisoning by his favorite reagents, the hydrazines. He had worked for fifteen years with "these dearly beloved bases, which were my first and most enduring chemical love affair," to use his own words. Fischer was not unaware of their dangerous nature, for many of his assistants had suffered from hydrazine poisoning, but he had considered himself immune to it. Now he became violently ill with symptoms of the gastrointestinal tract. The effects of the poisoning would be with him for twelve years at least; in fact, it is doubtful that he ever fully recovered.

Academic life in Erlangen apparently was not as congenial to Fischer as his time in Munich had been. When during the long convalescence from his difficult bronchitis, he learned that there was an opening as professor of chemistry in Würzburg, he was strongly interested. The Würzburg faculty had actually thought of him as its top candidate, but unfortunately the word got out that he was ill and another man was being considered for the position. When Fischer understood what was in the wind, he let it be known that he had fully recovered and was prepared to take on a new post. The faculty dispatched one of its leading members, the zoologist Semper, to interview Fischer and at the same time form an opinion about the state of his health. In his memoirs Fischer relates rather amusingly that the old professor took him for a long walk at a very brisk pace to test his physical fitness. Fortunately, Fischer was well enough to outpace his interviewer, who arrived at their destination panting and out of breath, completely convinced of the excellence of the candidate's health and stamina.

Fischer stayed in Würzburg for seven contented and productive years. He continued his work on carbohydrates and soon made a name for himself as a leader in the field. He pursued his investigations of the group of heterocyclic compounds that he in 1884 had termed purines. One member of that group, uric acid, had been discovered in 1776 by Scheele. A long row of distinguished chemists had worked on it. These included a colleague of Fischer's in Würzburg, Ludwig Medicus, who in 1875 had proposed a structure for uric acid as well as several other purines. Fischer's superior command of organic chemistry was needed, however, to definitively establish these structures by synthesis of the compounds, even though uric acid itself was synthesized in well-defined steps by Behrend and Roosen in 1888. It is a measure of the enormous amount of work put in by Fischer and his collaborators that at the turn of the century about 130 purines had been synthesized in his laboratory. Among the purines whose structure had been elucidated were adenine and guanine, which had been found to be present in both DNA and RNA (see Figure 1).

During the Würzburg years Fischer married Agnes Gerlach, the daughter of a distinguished anatomist in Erlangen. Fischer was thirty-six years old, and until then had not had much time for the ladies. His memoirs do not tell us much about his marriage, but it seems to have been a solid one and it was blessed with three sons, two of whom were born in Würzburg. Fischer had been very happy there, and when in the summer of 1892 he was approached by the authorities in Berlin and offered the prestigious chair in chemistry at the university, at first he was not interested. One of the reasons may have been that he was still suffering from the effects of the hydrazine poisoning. His wife intervened and persuaded him at least to go to Berlin and find out more about the position and the facilities.

When he arrived, he was not favorably impressed. He found Berlin ugly compared to Würzburg, and he considered the general atmosphere of the capital too stiff and Prussian for his taste. Since by this time he had a reputation as the leading organic chemist of the younger generation in Germany, the University of Berlin was ready to offer him almost anything, including a new chemical institute, in order to bring him there. When he finally accepted, their promises would turn out to be empty—or at least take a long time to fulfill. The new institute, for instance, did not materialize until 1900.

Fischer would remain in Berlin for the rest of his life, and in many ways his years there were extremely successful. With the reputation he had made for himself, it was easy for him to attract first-class collaborators. He soon built up a department that rapidly became internationally famous, if not legendary. He continued his pioneering research on carbohydrates and purines, and in 1902 received the Nobel Prize in Chemistry in recognition of these contributions. His work on amino acids and peptides has already been mentioned, and it is fair to say that he enjoyed a unique position as the leading chemist of his time. His brother-in-law, Max Friedrich, had certainly been proved wrong when he in 1869 declared that his unfortunate young apprentice would never amount to anything. Fischer's life was professionally successful to an extent that must have exceeded his most ambitious expectations; but his personal life was haunted by misfortunes.

After a marriage of seven years, his wife died suddenly of meningitis caused by a neglected ear infection. He was left with three small boys, of whom the youngest was only a year old. Fischer never remarried. From the time of his wife's death his household was managed by the faithful Margarete Barth, who remained with him until the end of his life. The eldest of Fischer's three sons, Hermann, became an outstanding biochemist, but the two younger boys came to sad ends.

The second son had decided to study medicine, but after two years in medical school suffered a nervous breakdown. When he had recovered, he was called up for military service as a doctor in 1914. After a few months in the service, he had a new breakdown and was put on sick leave. His illness progressed, interrupted by periods of relative health, until in 1916 he had to be admitted to a mental hospital. He became deeply depressed and eventually committed suicide at the age of twenty-five. In his autobiographic sketch Fischer gives a detailed account of his son's illness and death but neglects to tell us his name. It would seem that even two years after the boy's death, when Fischer wrote these notes that otherwise overflow with names, the pain was still so acute that, perhaps subconsciously, he could not bring himself to mention the name of this son.

The youngest son, Alfred, had wanted to become a chemist like his father and eldest brother, but Emil persuaded him to study medicine instead. The boy had a very sensitive skin, and Fischer had seen the effect of hydrazine on his own skin. While Alfred was studying at Heidelberg the war broke out. In 1915 he joined the army as a medical orderly. After a year he was promoted to assistant medical officer and sent to von Falkenhayn's army in Romania. In the spring of 1917 he was stationed in an isolation hospital in Bucharest, where the sanitary conditions seem to have been particularly bad. He was infected with typhus fever and died two weeks later. Hermann Fischer, who was serving on von Falkenhayn's staff, was given a short leave to attend the funeral of his youngest brother.

It is easy to understand why Emil Fischer in his memoirs repeatedly referred to the World War as *dieser unseliger Krieg* (this execrable war). He had never been a German chauvinist and in the days after the French defeat in the Franco-Prussian War, when Germany seemed to dominate continental Europe both economically and scientifically, Fischer always tried to promote scientific relations with France. There can be no doubt that the concept of science as an international brotherhood was something in which he believed strongly and that the war came as a shattering blow. When Fischer died in the summer of 1919, the world he had known lay in ruins around him. Perhaps the great chemist felt that it was time for him to go.

The Little Man from Russia

Anti-Semitism, one of the many dark chapters in the history of Europe, can be traced back to early medieval times and even to the Greco-Roman world. Its ancient roots are of course mainly religious, and the Catholic church has much to answer for in this regard. Islam was far more tolerant and in fact provided a haven for Judaism. In the nineteenth century anti-Semitism gradually changed its character and began to appear in a new, racist form. At the same time, it seemed to grow in strength and flared up in many countries of Central and Eastern Europe.

Nowhere was this evil more in evidence during this period than in Imperial Russia. In this country with its frightening heritage of anti-Semitism and pogroms, Phoebus Aaron Levene (Fedya to his friends) was born in 1869, the second child of Solom and Etta Levene in the little town of Sagor. The rather curious name Phoebus was not the one originally given him. That was Fishel, a Jewish name that was changed to the Russian Feodor when the family two years later moved to Saint Petersburg.

Solom Levene was a custom shirtmaker and seems to have prospered in the Russian capital, where he ultimately had three shops, one of them on the fashionable Nevsky Prospect, a main avenue in the city. The family grew until there were eight children. Since Solom was doing so well in the shirt-making business, he could afford to send Fedya to the Classical Gymnasium, where the emphasis, as the name implies, was on the teaching of Latin and Greek. Fedya became an accomplished linguist and later in life could read Newton's *Principia* in the original Latin. He also spoke French and German fluently, in addition to his rather heavily accented English, not to mention a smattering of Italian and Spanish.

After having graduated from the Gymnasium, Fedya passed the demanding examination for entrance to the Saint Petersburg Imperial Medical Academy and was admitted as one of the very limited number of Jewish students that the rules allowed. At the academy the famous composer Alexander Borodin was professor of chemistry, while his son-in-law Alexander Dianin was professor of organic chemistry. Levene attracted the attention of Dianin, who introduced him to chemical research and advised him to study chemistry further.

From then on Fedya seemed to lean more to basic research than to clinical medicine.

Before he had completed medical school, the increasing anti-Semitism in Russia had erupted into real pogroms, and in 1891 the Levene family decided to emigrate to the United States. It was now that Fedya changed his name to the unusual "Phoebus," which he mistakenly thought was the proper English rendering of Feodor. When a couple of years later he realized that a more appropriate English form would have been Theodore, he was already known by his new and rather original name and decided to keep it. It was as Phoebus Aaron Levene that he would be recognized the world over as one of the leading nucleic acid chemists.

Levene returned for a short while to Russia to complete his medical examinations. In the spring of 1892 he was back with his family in New York. After qualifying as a doctor there, he practiced medicine in the Russian Jewish colony until in 1896 he came down with tuberculosis. This illness necessitated a cure of about two years, which he spent first at Saranac Lake outside New York and then at Davos in Switzerland. When he had recuperated, he worked for a time in Albrecht Kossel's laboratory in Marburg and later with Emil Fischer in Berlin. This early contact with two of the leading laboratories in Europe must have confirmed his decision to devote himself to research rather than the practice of medicine. In 1905 he was recruited to the Rockefeller Institute for Medical Research; in 1907 he became a member of the Institute and was put in charge of its division of chemistry. He continued to work there until his death in 1940. Levene had become very fond of Saranac Lake and often returned there for a vacation. It was on one of these trips that he met Anna Erickson, a woman of Norwegian ancestry, whom he married in 1920.

Phoebus Levene is described as a thin, wiry man of short stature with penetrating, dark eyes and a shock of black hair that later turned steely gray. Intense and lively with a somewhat stern, perhaps even authoritarian, aspect, he was a glutton for work. His whole life centered on his laboratory at the Rockefeller Institute and his habits were exceedingly regular. Although he traveled widely in Europe, his knowledge of the United States scarcely extended beyond New York; anything west of the Hudson River was unknown territory. In spite of his seemingly complete preoccupation with science, he was widely read and his apartment overflowed with books. He was particularly interested in art and artists, having rather modern taste. At home his walls, where not occupied by bookshelves, were covered with paintings. His social life included not only scientists but also a number of artists and literary people.

Some of Levene's early contributions to nucleic acid chemistry have already been mentioned, but his main discoveries concern the structure of the nucleotides and the way these building blocks of nucleic acids are linked to one another. As early as 1847, Justus von Liebig had isolated an acid from beef muscle that he called inosinic acid. Almost fifty years later, the German biochemist Franz Haiser had confirmed Liebig's discovery and shown that inosinic acid contained hypoxanthine and phosphate. In addition, on acid hydrolysis it yielded yet another compound that was later identified as a five-carbon sugar, a pentose. Much confusion had existed regarding the nature of this pentose until Levene and his collaborator Walter Jacobs in 1909 established unequivocally that it was D-ribose. They were also able to show that in inosinic acid the ribose was linked to hypoxanthine by what is now called a glycosidic bond, while the phosphate moiety was esterified on one of the sugar hydroxyls (see Figure 1).

From inosinic acid the two men turned to a substance that Olof Hammarsten had isolated after alkaline hydrolysis of a "nucleoprotein" obtained from pancreas and given the name guanylic acid. He had demonstrated that it contained phosphorus and yielded guanine on hydrolysis, hence the name. In 1907 guanylic acid was shown by Kossel's associate Hermann Steudel to be made up of guanine, pentose, and phosphate. Levene and Jacobs now found that these components of guanylic acid were linked to each other in the same way as were hypoxanthine, ribose, and phosphate in inosinic acid, and they also identified the pentose as ribose. They introduced the term "nucleotide" to denote a compound containing a purine or a pyrimidine (often referred to as a "base") linked to ribose and phosphate, whereas base and ribose alone without phosphate was called a "nucleoside."

Having established the general structure of inosinic acid and guanylic acid, Levene and Jacobs proceeded to analyze the products obtained after alkaline hydrolysis of yeast nucleic acid, using the same procedure that had yielded nucleosides from nucleotides by cleavage of the ester bond between phosphate and the sugar moiety. They obtained four different nucleosides in what appeared to be roughly equimolar proportions; each of these nucleosides on acid hydrolysis yielded one of the following bases: adenine, guanine, cytosine, or uracil. From these results Levene and Jacobs concluded that yeast nucleic acid was made up of adenylic acid, guanylic acid, cytidylic acid, and uridylic acid in the form of a tetranucleotide.

It should be pointed out that the methods available at the time were far from quantitative. Furthermore, it was not recognized that the samples of nucleic acid analyzed were not homogeneous entities, but mixtures of a great number of individual RNA molecules with varying base composition. The apparent equimolar proportions of the four nucleotides were therefore fortuitous. Levene, of course, did not know this and to the end of his days firmly believed that both RNA and DNA were tetranucleotides—or, at the very most, represented a few repetitions of this structural motif. His tetranucleotide hypothesis, widely accepted during the first half of this century, thus underestimated in a grotesque way the gigantic size of these macromolecules. It did a lot of harm in the sense that it diverted people from the correct idea of the late nineteenth century, that nucleic acids are the carriers of genetic information.

Levene's last and perhaps most significant contributions to nucleic acid chemistry were his elucidation of the structure of deoxyribose, and his conclusion in 1935 that in DNA the nucleotides are linked to one another by way of phosphodiester bridges, in which the phosphoric acid is doubly esterified to carbon atom 3′ of one deoxyribose and carbon atom 5′ of the sugar of the adjacent nucleotide (see Figure 1). Almost twenty more years would pass before it was established that the same general arrangement is true also of RNA.

Alexander Todd, later Lord Todd of Trumpington, was a dominating figure in nucleic acid chemistry with an almost Medicean magnificence of personal appearance. Not least among his achievements was the unequivocal establishment through chemical synthesis of the nature of the linkage between sugar and ring nitrogen in the nucleotides. The synthetic methods developed in his laboratory permitted the synthesis of artificial substrates for the RNA-degrading enzyme ribonuclease. By analysis of the cleavage products obtained with this enzyme, it was definitely established in 1952 that RNA is made up of nucleotide units that are linked to one another through phosphoric acid, doubly esterified to carbon atom 3′ in the ribose moiety of one nucleotide and carbon atom 5′ of the adjacent nucleotide (see Figure 1). This solution of an important problem in RNA structure also reinforced Levene's tentative structure of DNA—a necessary prerequisite for the double helix proposed by Watson and Crick a year later.

There can be no doubt that the name of Phoebus Levene will forever be linked to some of the decisive discoveries in nucleic acid chemistry. At the same time, his unchallanged authority and perhaps also his dominant personality helped to perpetuate some unfortunate misconceptions that would persist for a long time.

The tetranucleotide hypothesis rested on two main points: the fortuitous observation of equimolar proportions of the four nucleotides when undefined mixtures of RNAs were analyzed, and the complete lack of reliable information about the molecular weight of either DNA or RNA. A few previous reports had indicated that nucleic acids might be what today we call macromolecules. Friedrich Miescher, for instance, had found that his salmon sperm nuclein did not diffuse through a membrane permeable to smaller molecules, but these early observations were disregarded or put down to aggregation of the ubiquitous tetranucleotides.

It was therefore an important step forward when Olof Hammarsten's nephew, Einar Hammarsten, and his collaborators could show in 1938 that DNA carefully prepared from thymus behaved like thin, rodlike molecules with an apparent molecular weight that approached a million. Even higher molecular weights were later reported by John Gulland but not until the middle of the century did the idea that DNA, and perhaps also RNA, was a true macromolecule become widely accepted.

The canonical principle that nucleic acids contained equimolar proportions of the four standard nucleotides was shattered when Erwin Chargaff began to analyze the base composition of DNA from many different sources, using the new chromatographic techniques that had become available. In 1950 he could report that DNA showed a phenomenon called base complementarity. That is, it contained equal amounts of guanine and cytosine, and of adenine and thymine. On the other hand, the ratio of (guanine + cytosine) to (adenine + thymine) varied widely for DNAs from different sources. Consequently, the tetranucleotide hypothesis was untenable, since DNA did not contain equimolar proportions of the four nucleotides.

With the demise of this unfortunate hypothesis and the recognition that DNA was a macromolecule, one of the main reasons for the rejection of DNA as the molecular basis of genetic information had finally disappeared. But even when definite proof of DNA's role in heredity was presented in the 1940s, it still met with considerable resistance.

[five] The Dawn of Molecular Genetics

The practical application of empirical genetics, completely without a theoretical foundation, is as old as farming and stock raising, when humankind first learned to select among crops and domestic animals the genetic varieties best suited to its own purposes. On the other hand, genetics as a science is comparatively young and has its roots in a corner of the garden of the Augustinian monastery at Brünn.

The Catholic church can hardly be said to have lacked interest in cultural matters during its almost two-thousand-year history. On the contrary, whenever the church was not preoccupied with matters of greater importance (like upholding its supremacy against the insidious and godless machinations of the German emperor or suppressing heretics by burning them at the stake), it protected European art and humanism. Without it our cultural inheritance would be incomparably poorer. But it was only one of the two cultures, to use C. P. Snow's terminology, that was encouraged and patronized—the one represented by art and the humanities. As far as science was concerned, the church was at best uninterested, and often actively hostile. It is enough to recall such names as Giordano Bruno and Galileo Galilei to substantiate this claim. So it is remarkable that a prince of the Catholic church, albeit one of the least prominent and pretentious, must be credited with having fathered scientific genetics, one of the outstanding achievements in science and the foundation of modern biology.

The Concept of the Gene

Johann Mendel was born in 1822 of Silesian peasant stock. Like many other talented but impecunious young men over the centuries, he looked to the Catholic church for a career. At the age of twenty-one he was admitted as a novice in the Augustinian monastery at Brünn. On this occasion, following an old tradition, he received a new Christian name, Gregor. It was the name by which he would, long after his death in 1884, be known all over the scientific world. He had had some previous training in science, and fortunately his abbot recognized his unusual talents and sent him to the University of Vienna (at the time Silesia was part of the Austrian Empire) to continue his science studies. After three years Mendel returned to the monastery in 1854. He was employed as a science teacher by the technical high school in Brünn and in 1856 began in his spare time to perform genetic experiments with plants that he grew in the monastery's garden.

He chose to work with the garden pea and such easily observable genetic variations (phenotypes) as tallness and dwarfism, color or absence of color in blossoms, seed shape, and the like, where each character had remained constant for generations. When Mendel crossed one variety, characterized by tallness, with a dwarf variety, he found that all the plants of the first hybrid generation (F1) were tall; there were no dwarfs. He called the expressed character (tallness) dominant and the suppressed character (dwarfism) recessive. When the first hybrid generation produced a second generation (F2) by self-fertilization, both the dominant and the recessive characters could be observed in the offspring, and they appeared in the same constant proportion in all experiments: three-fourths of the plants were tall and one-fourth were dwarfs. When he tested the F2 generation individually, he found that the dwarf plants on self-fertilization gave exclusively dwarfs in generation after generation; they were what we call homozygotes. Of the tall F2 plants, one-third gave only tall offspring, they were homozygotes for tallness, while two-thirds resembled the F1 generation and gave three-fourths tall offspring and one-fourth dwarfs; they were heterozygotes (Figure 2).

From these and similar results Mendel concluded that the genetic characters or units, which would later be called genes (for instance, those representing tallness and dwarfism), always occurred in pairs. The homozygous parent generation would have the characters AA (tall) and aa (dwarf). The mature germ cells (gametes) of the parent generation would contain one unit each, either A or a. The hybrid F1 generation would have the characters Aa, and in the gametes of that gen-

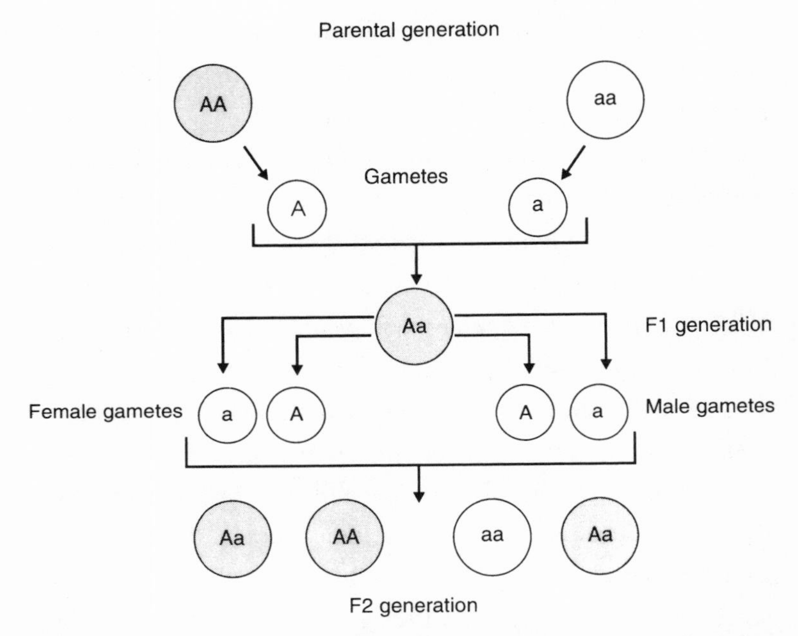

Figure 2 Mendelian inheritance: segregation in the gametes from the F1 generation of the alternative characters A (dominant) and a (recessive), according to Mendel's first law. Individuals whose observable properties (phenotypes) correspond to the dominant character A are shown as filled circles; individuals with phenotypes corresponding to the recessive character a are shown as open circles.

eration there would be a segregation of those characters. This principle is called Mendel's first law.

Mendel went on to study several different pairs of characters and found that they were independently transmitted to the succeeding generations. This so-called independent segregation is referred to as Mendel's second law. As will be apparent, this principle is not universally applicable. As a rule it holds only for genes on different chromosomes.

What is so remarkable about Gregor Mendel's work is the systematic way in which he analyzed and quantified the results of his experiments and understood the importance of statistical methods in genetics. It has been claimed in recent years that his results were too good to be true, that in fact he may have selected his experimental data to make them agree better with his preconceived ideas. Be that as it may, it in no way detracts from the conceptual importance of his seminal work. It is inconceivable that an upright and honest man such as Mendel would deliberately commit an act of what he himself considered fraud. Still, opinions on what is and what is not acceptable in the representation of biological results may have changed appreciably over more than a century; in Mendel's time the criteria were probably less strict.

In 1865, when Mendel for the first time presented his results at a meeting of the Natural Science Society in Brünn, he could do so with a certain amount of self-confidence. This assurance is obvious also in the publication *Versuche über Pflanzen-Hybriden* (Experiments with plant hybrids), which appeared the next year in the transactions of the society. Here his work remained buried, until in 1900 it was resurrected by three botanists—Hugo de Vries, Karl Correns, and Erich Tschermack von Seysenegg—who, independently of Mendel, had arrived at similar conclusions. One must salute the thoroughness of their review of the literature, which brought to light Mendel's paper in this somewhat obscure journal of thirty-four years previous. Even more impressive is the sense of honor that caused them, at their own expense, to elevate Gregor Mendel to world fame, instead of absentmindedly letting his work sink into anonymity and oblivion.

Sadly, for the rest of his life Mendel had very little time for science. In 1868 he was elected abbot of his monastery, a high and responsible office within the church that brought with it a great deal of administrative work, not to mention a long and acrimonious conflict with the Austrian authorities over taxation of the monastery. Surely he would have been much happier had he been allowed to continue his research in a peaceful corner of the monastery garden.

The rediscovery of Mendel's laws brought about a period of prosperity and rapid growth for the new science. It was beginning to be called genetics, in accordance with a suggestion of Wilhelm Johannsen that the units of heredity should be termed genes. At this point something rather unexpected occurred. Those of us who work in the basic sciences related to medicine take every opportunity to emphasize the enormous importance of such research as the basis for progress in the diagnosis and treatment of practically all illnesses. Yet we tend to forget the many instances when a clinical observation has put, for instance, biochemists or physiologists on the right track.

A classic example is the British physician Archibald Garrod (1857–1936) and his discovery of "inborn errors of metabolism," the title of the book published in 1909 in which he introduced this new concept. While working as a doctor in a London hospital, Garrod had become interested in various pigments formed as a result of pathological processes and excreted by the patient in the urine. He realized that in order to pursue this interest he needed to become better grounded in biochemistry. To that end he collaborated with Britain's leading biochemist at the turn of the century, Frederick Gowland Hopkins, affectionately known as Hoppy. He then felt able to tackle a condition known as black diaper disease, or alkaptonuria, observed in newborn babies. The characteristic finding was that the urine of these infants turned black when exposed to air, hence the name of the disease. He realized that the black pigment originated from the accumulation of a breakdown product of the aromatic amino acids phenylalanine and tyrosine, previously identified as homogentisic acid.

When Garrod studied the heredity of this relatively benign condition, he found that consanguinity among the parents seemed to increase the frequency of alkaptonuria in the offspring. William Bateson, a botanist and geneticist, suggested to him that the condition was transmitted according to Mendel's principles for a recessive character, and Garrod readily accepted that explanation. It should be emphasized that Garrod published his observations in 1902, only two years after the resurrection of Mendel's laws—no mean achievement for a clinician without any genetic training. In his Croonian lectures, "Inborn Errors of Metabolism," delivered in 1908 and published the next year, came his great conceptual breakthrough. He boldly suggested that the genetic defect in alkaptonuria was the absence of a certain enzyme necessary for the normal breakdown of tyrosine and that this lack led to the accumulation of homogentisic acid. For the first time, the connection between a gene and its gene product (the enzyme) could be divined; at the same time a new group of diseases, the hereditary metabolic conditions, made their appearance on the medical scene.

Garrod was far ahead of his time, and no one seems to have paid much attention to this new concept, which portended the discovery by George Beadle and Edward Tatum more than thirty years later of the relationship between gene and gene product. The medical Nobel committee, to its embarrassment, never quite realized what a decisive contribution Archibald Garrod had made.

During the first decades of the twentieth century, genetic research was dominated by Thomas Hunt Morgan and his collaborators at Columbia University. In 1903 Walter Sutton realized that genes were associated with chromosomes and that each chromosome contained several genes. A number of exceptions to Mendel's principle of independent segregation of genetic characters were also discovered. Based on his extensive and thorough investigation of sex-linked genetic factors in the vinegar fly, *Drosophila melanogaster*, Morgan established that genes appear in a linear arrangement on the chromosome. Thus, genes on the same chromosome are linked to one another to form a so-called linkage group. This concept explained the exceptions to Mendel's second law of independent segregation: as a rule, only genes belonging to different linkage groups can segregate independently.

Of fundamental importance to his concept of genes and chromosomes were Morgan's studies of the exchange of genes between homologous chromosomes during meiosis (the formation of the mature gametes). This phenomenon he called crossing over, and he used it to determine the distances between and relative order of genes on the chromosome. The principle of the method originally proposed by Alfred Sturtevant was that the farther apart genes were, the more likely it was that a crossing over during meiosis would lead to an exchange of such genes between homologous chromosomes.

The pioneering work of Thomas Hunt Morgan and his collaborators (Sturtevant and Hermann Muller, for instance) created a firm experimental basis for the new science and at the same time made *Drosophila* extremely popular among geneticists. In the mid-1930s the American George Beadle and the Russian-born French biologist Boris Ephrussi also worked with *Drosophila.* Studying mutants with changes in the normal brilliant red color of the eyes, they demonstrated that the various chemical steps that led to the formation of the eye pigment were controlled by separate genes. Thus, these genes had to give rise to the different enzymes necessary for biosynthesis of the pigment.

As an experimental system to be used in an attempt to relate genetics to biochemistry, *Drosophila* was too complicated; a simpler organism was needed. Beadle realized this, and together with his new collaborator Edward Tatum he chose to work with the mold *Neurospora*. Here was an organism whose genetics had been fairly well worked out and that had quite simple nutritional requirements. The two men produced a number of mutants of *Neurospora* by irradiation with X-rays, a method first used by Hermann Muller when working with *Drosophila*. Among the mutants they found one that had lost its ability to synthesize the vitamin pyridoxine and that therefore required the addition of this vitamin as a nutrient for growth. By systematically testing pyridoxine-requiring mutants in terms of their enzyme deficiencies, in 1941 they arrived at the important hypothesis "one gene, one enzyme"—or to put it in more general terms, "one gene, one protein." In other words, for every protein in the cell (or rather, for every polypeptide, since many proteins are made up of several different polypeptide chains) there is a gene that contains the structural information necessary to produce that polypeptide.

By the 1940s scientists had come a long way in terms of understanding the function of the gene, but its chemical nature remained an enigma. Most biologists took the view that genes must be proteins. These complicated giant molecules were already known to have sophisticated functions as biological catalysts; why could they not be carriers of the genetic information as well?

An Old Bachelor in New York

Today the United States dominates biomedical research, a fact that is vividly demonstrated every December 10, when the King of Sweden hands out the Nobel Prizes. One hundred years ago the situation was radically different. At the end of the nineteenth century basic medical research was concentrated in Europe and the United States had virtually no role at all. American medicine seemed destined to plod along in its usual way without being able to contribute anything to the spectacular progress in basic medical sciences being made in European laboratories.

At this point the man of destiny enters the scene, in the form of the Baptist pastor Frederick T. Gates. An altogether unlikely figure, one might think. But the Reverend Mr. Gates was spiritual adviser to John D. Rockefeller, one of the robber barons who had made incredible fortunes at the end of the century. At this time William Osler was the dominant figure in American medicine. He was involved in purely clinical observations, what we call bedside medicine, and had no interest whatever in experimental biomedical research. At the same time, he was an outstanding clinician with a healthy skepticism of the often-meaningless treatments employed in contemporary medicine. This critical attitude toward the conventional therapeutic wisdom, always the hallmark of a good clinician, was very much apparent in Osler's famous textbook, *The Principles and Practice of Medicine.*

Now an unlikely event occurs. The Reverend Mr. Gates, always interested in medical questions, sits down with Osler's mighty tome and reads it cover to cover. The thought of a present-day pastor doing something of this sort is almost inconceivable. But in his earlier experience as a priest working among the common people, Gates had seen too many examples of how helpless medicine was when confronted with serious illness. With a spirit of down-to-earth Christianity, Gates decided to do something. Christ had driven the money changers out of the temple and said that it was easier for a camel to pass through the eye of a needle than for a rich man to enter heaven. Gates was probably not prepared to go that far. Instead, he befriended one of the wealthy tycoons and persuaded him to give generously to various charities—in this way perhaps facilitating his entry into heaven, when the matter of that hoped-for passage would eventually arise.

One might surmise that the Reverend Mr. Gates would try to talk Mr. Rockefeller into donating money for a fine new hospital of the traditional type. Not at all. What he wanted was something completely new to America, an institute aimed exclusively at biomedical research, where the hospital that was part of the institute existed only to sustain clinical investigations. Gates perceived clearly that real clinical progress was impossible without the basic medical sciences. One must admire the old Baptist pastor who one hundred years ago saw farther than most European politicians today!

Thus was the Rockefeller Institute founded in 1901, creating unique opportunities for science-oriented biomedical research in the United States. The difference in attitude became apparent also in the choice of scientists for top positions in the new institute. Perhaps the most colorful among them was the German physiologist Jacques Loeb, noted for the militant, even fanatical, way in which he asserted the self-evident supremacy of medical science over traditional clinical medicine. His fame reached even the literary field: Sinclair Lewis, in his well-known novel *Arrowsmith*, made his young doctor-hero an ardent admirer of Loeb.

In 1913 these most splendid paladins of medical research in the United States were joined by an individual who in every respect seemed to be the antithesis of Jacques Loeb, Oswald Theodore Avery (1877–1955). He was very small and slender, his face was dominated by a high forehead, and his whole appearance was reminiscent of a science-fiction writer's vision of man in the distant future: the brain seemed to have taken over completely, while the body had been reduced to a minimum.

Avery was a physician who had tired of the clinic and wanted now to devote himself entirely to research. From the day he came to the Rockefeller Institute until the day he retired at the age of seventy-one, his life was dedicated to the study of *pneumococci*, a type of bacteria that can cause pneumonia. Before the advent of sulphonamides and penicillin, this disease was a scourge that claimed innumerable victims each year. It was only natural that the Rockefeller Institute should make pneumonia the principal object of its experimental clinical research.

When Avery decided to devote his life to the elucidation of the secrets of the *pneumococcus*, the sentimental cliché about heroic doctors was for once literally true. We run-of-the-mill scientists tend to complicate our lives in a way that is completely indefensible from the point of view of science. We enter into complicated relationships, marry and divorce, bring children into the world, build ourselves a home or two, buy cars, boats, and the countless other wordly goods that are part of the usual domestic life. We sit in academies and on research councils, join advisory committees, and write evaluations of every conceivable subject. In short, we waste precious time every day, time that should instead be devoted to research.

Avery did none of these things. A confirmed bachelor, he lived within walking distance of his laboratory, did not travel abroad unless forced to, hardly ever participated in social life, and never seems to have fallen victim to any temptation of life in New York. His only weaknesses appear to have been a fondness acquired in early youth for playing the cornet and a passion for sailing picked up much later in life. Even then he would spend his time philosophizing in the cockpit while others attended to the practical details of sailing. Avery was extremely wary of public appearances and hated to give lectures. When he actually had to give one, it was a stylistic masterpiece and thoroughly rehearsed beforehand.

This description may give the impression that Avery was a hopeless recluse without any contact with other human beings; but that was not the case at all. His need for a social life was fully satisfied during his working hours at the institute. He had many friends there and was beloved by his younger collaborators and by the senior members of the Rockefeller Institute. The little man, so shy when he appeared in public, was fond of talking for hours with his "boys," as he called them, and displayed a flowing eloquence that developed into long monologues in which he expounded his scientific theories.

During the period between wars Avery achieved international fame as a scientist, based primarily on his investigations of the capsule that surrounds virulent *pneumococci*. Together with Michael Heidelberger, he showed that the capsule was made up of high-molecular-weight carbohydrates (polysaccharides) and that these determined the ability of the bacteria to cause the formation of antibodies in the infected organism that were directed against *pneumococci*. He was also able to use polysaccharides in combination with proteins to produce sera that proved useful for treating patients with pneumonia. However, his greatest discovery was still to come. It would open up a whole new field of research, that of molecular genetics.

That story begins in the 1920s and on the other side of the Atlantic. In the pathology laboratories of the British Ministry of Health in London, the bacteriologist Fred Griffith was working with *pneumococci*. Like Avery, he was a quiet, retiring bachelor, small and slender, with a voice that was seldom raised above a whisper. Curiously this prim and correct public servant had a summer house built near Brighton in an almost shockingly modernistic style, and he loved to drive his car at breakneck speed.

Griffith's passion, though, was *pneumococci*. He discovered, as had Fred Neufeld in Germany, that there are two different forms of these bacteria, called S-*pneumococci* and R-*pneumococci*. The S-form, which is virulent and causes disease, has a capsule; the harmless R-form does not. In 1928 Griffith published a series of experiments that would have consequences for medical research that he could never have imagined.

Virulent *pneumococci* are dangerous for humans, as a long and depressing clinical experience had proved. They are even more dangerous to mice, in which an infection has 100 percent mortality, provided the *pneumococci* are of the S-type. On the other hand, mice can be infected with R-*pneumococci* without any ill effects.

Griffith devised an experiment in which he mixed living but harmless R-*pneumococci* with S-*pneumococci*, which he had killed by heating them to 60° C. He then injected this mixture into a series of mice. The experiment seems to have been prompted by his observation that the sputum of patients with pneumonia often contained several different serological types of *pneumococci* and his intuitive belief that this fact could not be explained by simultaneous infections with the different types. Instead, he believed that *pneumococci* could change from one serological type to another in the infected organism.

It would have been reasonable to expect the mice to survive the experiment, since there were no living, virulent *pneumococci* in the injected mixture. Instead, several of them died. Even more surprising, Griffith could grow virulent S-*pneumococci* from the blood of the dead mice. One might almost believe that the heat-killed S-*pneumococci* that he had injected into the mice had been resurrected. Griffith rightly rejected this theologically pleasing but scientifically unsatisfactory explanation. Another somewhat embarrassing possibility was that he had after all injected some living S-*pneumococci* into his mice, but in a series of careful control experiments Griffith satisfied himself that this was not the case. Living R-*pneumococci* must have taken up something from the dead S-*pneumococci* that transformed them into S-*pneumococci*, with all the characteristics of that form. What had occurred was a genetically stable change of form, the equivalent of what we today call a mutation, but in this case the term "transformation" is preferable.

Although Griffith certainly realized that his discovery was important, he never did really understand its implications. Unfortunately, he did not live to see the incredible development that resulted. Griffith was killed during a 1941 London air raid while serving as a volunteer firefighter—one of the countless victims of World War II. Many years later his name would emerge from obscurity and enjoy a celebrity that he could surely never have imagined during his lifetime. Given his retiring nature, he probably would have strongly disapproved.

Griffith's discovery aroused a certain amount of interest internationally, and in Avery's laboratory his experiments were repeated and the results confirmed. One of Avery's collaborators showed in 1932 that R-*pneumococci* could be transformed with an extract of S-*pneumococci*, technically a crucial step forward. Unlike Griffith, Avery realized that the transforming agent that could be extracted from S-*pneumococci* and that changed R-form into S-form must be the molecule that contained the genetic information, the genome, of the cell. If the transforming principle could be isolated and characterized, one of the most perplexing problems of biology—the molecular basis of heredity—would have been solved. For ten years Avery worked on this problem, experiencing enormous experimental difficulties and constant frustration. It must have been during this time that he coined his famous phrase, "Disappointment is my daily bread, but I thrive on it."

A major problem was that the extracts varied so enormously in transformation activity; often they were completely inactive. It was not until Avery's collaborator Colin MacLeod was able to devise a better extraction method and a more reliable way of measuring transformation activity that they could make real progress. In the summer of 1941 MacLeod became professor of bacteriology at the New York University and consequently disappeared from the institute, even though he frequently visited the laboratory and kept in constant touch with the project. His successor was Maclyn McCarty, a young pediatrician with extensive biochemical training. In a surprisingly short time he was able to purify the transforming principle and show that it was identical with one of the two known types of nucleic acids, DNA. This conclusion was supported by his finding that the transforming activity was destroyed by DNase, a DNA-splitting enzyme.

In these enlightened days, when every schoolchild knows that DNA is the bearer of the genetic information, Avery's conclusion seems rather self-evident and unremarkable. At the time of his discovery it was far from obvious. On the contrary, his claim that DNA was the transforming principle was met with deep suspicion in many quarters, in spite of the fact that Avery, MacLeod, and McCarty in their now-classic publication of 1944 expressed themselves with almost excessive caution—or perhaps for that very reason.

An embarrassing circumstance was the fact that P. A. Levene, as we have seen, had been of the opinion that DNA was made up of a monotonous repetition of its nucleotide components A, C, G, and T. Consequently it could not contain all the information present in the genes. Levene also, in an almost grotesque way, underestimated the size of the DNA molecule, which of course did not help Avery's case. DNA, a comparatively small and structurally simple molecule, the bearer of the genome? Absolutely out of the question! Proteins, on the other hand, known to be complicated giant molecules with sophisticated functions as biological catalysts, were a different proposition. Why should not proteins, in addition to all their other functions in the cell, contain the genetic information? The well-known biochemist Alfred Mirsky, a colleague at the Rockefeller Institute, was firmly convinced that Avery's DNA preparations contained protein, the true transforming molecule.

In 1952 Alfred Hershey and Martha Chase, using an entirely different biological system and experimental approach, obtained results that strongly supported Avery's general conclusions. They showed that when coliphage, its DNA labeled with radioactive phosphorus and its protein with radioactive sulfur, infected its host, the gut bacterium *Escherichia coli* (or *E. coli*), the phage DNA entered the host cell while the protein remained outside. Thus, DNA was responsible both for the transformation of *pneumococci* to give them new genetic properties, and for the ability of a bacteriophage to infect its host. Even confirmed skeptics like Mirsky were beginning to change their minds when faced with so much evidence from independent sources.

It had taken a comparatively long time before it was generally accepted that DNA could indeed transform *pneumococci*, and an even longer time before the scientific community fully realized what that meant: the DNA molecule contained the genes. To a certain extent one could say that this was Avery's own fault. After all, he was *so* extremely cautious. In his famous paper of 1944 he barely mentioned the word "gene" and he always worried about the small amounts of protein that his DNA preparations undoubtedly contained, regardless of how carefully he purified them.

A principal factor was surely his hatred of public relations and self-advertisement. When the Royal Society awarded him the prestigious Copley Medal in 1946, he refused to go to London for the awards ceremony on the pretext that because of his delicate health he could only travel first class, which would be far too expensive. As far as we know, he was in excellent health at the time and his finances gave no cause for concern, in spite of the fact that he had helped a younger brother through medical school and still supported an impecunious female relative. Most scientists would rise from their deathbed to receive the Copley Medal. Avery seems to have been completely indifferent to both fame and reward. The fact that he was never awarded the Nobel Prize was probably of no concern to him. In fact, there is reason to believe that all the attendant commotion would have been repugnant to him.

The fact remains that this gentle, quiet, little man, always very neat and as discretely well dressed as an old-fashioned family doctor, is one of the towering figures of biomedical research. Even if both James Watson and Francis Crick can claim with equal justification to be the father of molecular genetics, there can be no doubt that the grandfather was Oswald T. Avery!

Deus ex Machina in Cambridge

In my country we tend to become more and more settled and less and less mobile, so that nowadays a Swedish scientist cannot be persuaded to move from his or her alma mater to another university unless a professorship is the bait. Young American scientists have an entirely different attitude. Possibly this has to do with the tradition of "up and away" inherited from the pioneers, but U.S. scientists are much more inclined to cut loose and move to new frontiers than their European colleagues. After their doctorate comes a long postdoctoral period, working a couple of years at one university and then moving on to another, either to a new postdoctoral position or, God willing, to an untenured assistant professorship. Before striking camp and cramming the family and its few belongings into the car, there is often one of the garage sales so characteristic of American campuses, where everything that cannot be transported in the car is unsentimentally sold. A

Gypsy life, no doubt, but at the same time some stimulating experiences that really broaden one's intellectual outlook.

Almost half a century ago a young American postdoctoral fellow arrived in Copenhagen, where he was to work in the laboratory of the Danish biochemist Herman Kalckar. His name was James Watson, and his training to date had been mainly in microbiology and genetics. The idea was that he should now learn some biochemistry with Kalckar. On the face of it, his choice of laboratory seemed very reasonable.

However, Kalckar's life at this point was exceedingly complicated and he had little time for Watson. Furthermore, their exchange of thoughts was made difficult by the fact that Kalckar's English was fluent but very Danish in character. In his brilliant book *The Double Helix*, Watson tells a highly amusing story of how he used to bicycle home from these charming but incomprehensible conversations with Kalckar in a state of deepest frustration. Finally he decided to do something radical in order to escape his hopeless situation. He elected to go to Cambridge, England, in order to work in the Cavendish Laboratory. It was led by Sir Lawrence Bragg, that grand old man, together with his father, William Bragg, the virtual inventor of X-ray crystallography. After some administrative mixups he managed to get his postdoctoral stipend transferred from Copenhagen to Cambridge. Thereby he had unknowingly taken the first step on his way to world fame and a Nobel Prize.

When Jim Watson arrived at the Cavendish in the fall of 1951, he encountered a somewhat overage graduate student by the name of Francis Crick, whose scientific career had been interrupted by World War II. Crick was now doing his thesis work on the X-ray crystallographic analysis of proteins. His background was mainly that of a physicist, but his passion could best be described as theoretical biology, particularly in the fields of biochemistry and biophysics. In certain kinds of research the work tends to be decidedly practical and experimental, and the incentives for elegant theories are rather meager. Biochemistry is a case in point. It is easy to imagine that a theoretical physicist could perform scientific miracles armed only with his own genius (and in more recent days also a tremendously expensive supercomputer). A biochemist, on the other hand, should work at the bench with his test tubes, should he not? If one wanted to be unkind, one might say that we biochemists and molecular biologists are not grand, individualistic thinkers. Rather, there seems to be a general consensus on which way the field is moving and what ought to be done, so it is a question of getting it done as quickly as possible. But there are no rules without exceptions, and Crick is our brilliant exception to this rule.

A young Austrian scientist named Max Perutz came to the Cavendish in 1936 and managed to stay on there, which was fortunate indeed since he is of Jewish ancestry and in this way could escape from Hitler and his henchmen. He became one of the leaders in protein crystallography (and was incidentally the thesis adviser of Francis Crick). While Perutz is an extremely modest and retiring person, he is nevertheless one of the finest scientists of our time.

It is difficult to imagine two more disparate characters than Crick and his thesis adviser. The very thought that Crick ever had an adviser boggles the mind—he who has always seen farthest and known best! This was already the case when he was just a greenhorn in science. Francis Crick has never been one to hide his light under a bushel or refrain from criticism, regardless of how illustrious the victim was. In fact, Watson begins his book with the statement, "I have never seen Francis Crick in a modest mood." On the whole, one is inclined to believe him.

Not even such a national monument as Sir Lawrence Bragg escaped Crick's sharp tongue. He thought nothing of challenging Sir Lawrence's opinions about the proper way to do protein crystallography, even in the presence of the great man himself. One can understand that Crick, at least at the beginning of his career at the Cavendish, was not one of Bragg's favorites. What especially irritated Sir Lawrence was Crick's habit of loudly criticizing the lecturer during a seminar. Since Crick has a voice that carries over considerable distance, the unfortunate speaker could be in no doubt about his intellectual deficiency. On one such occasion Crick happened to be sitting in the audience right behind Sir Lawrence. Finally Bragg's patience was exhausted and he turned on Crick with what must now be a historic admonition: "Crick, you're rocking the boat!"

Rocking the boat is exactly what Crick has been doing his whole life, and with great success. That is not to say that he is brutal and insensitive (there are such people even in science).

On the contrary, he is very much the British gentleman; but he may become a bit excited and perhaps even irritated when people in his opinion are unpardonably dense and slow on the uptake.

When fate, in its somewhat absentminded and capricious way, brought him together with Jim Watson, freshly escaped from Copenhagen, it created one of those uniquely effective teams in which the parties mysteriously seem to complement and stimulate each other. Their originality and complete lack of respect for established authority and conventional wisdom would revolutionize biology.

The problem that fascinated Watson and Crick and took up most of their time and interest (officially they were supposed to be hard at work on entirely different projects) was the three-dimensional structure of DNA. It had taken the scientific community a surprisingly long time to accept Avery's discovery that DNA is the molecular basis of heredity, but its time seemed finally to have come. Even a chemist of the stature of Linus Pauling had become interested in DNA structure. By careful model building based on an intimate knowledge of the structure of amino acids, the building blocks of proteins, he had been able to implicate such structural elements as the alpha helix in proteins.

The idea of helices in biological macromolecules was therefore in the air, and it was only natural that Watson and Crick had it in mind when they tried to tackle DNA structure. They had learned from Pauling that by knowing the structure of the building blocks, the distance between the atoms, and the angles between the bonds that held them together, one could build meaningful molecular models. A prerequisite was knowing enough about the structure of the building blocks to control the three-dimensional fantasy and prevent building just any kind of model. The more structural restrictions there were, the more exact and meaningful the model would be. The lack of such restrictions was the main problem in their DNA model building. At this point an imposing yet somewhat tragic figure enters the scene.

It would seem a natural conclusion that cultural prosperity coincides with and presupposes periods of political stability and success in a country. However, this putative connection is doubtful indeed, and an example is the monarchy of Austria-Hungary at the turn of the century. It is difficult to imagine anything more politically unstable and full of disastrous internal tensions than the conglomerate of seemingly incompatible nationalities and cultures that made up the proud Hapsburg realm rooted in the Holy Roman Empire of the German nation.

The only cement that seemed to hold this creaking edifice together and prevent it from falling into ruins was the elderly emperor himself, Franz Joseph. Early every morning he sat down at his desk with a kind of hopeless faithfulness, to deal with the endless stream of administrative items produced by an inefficient and corrupt bureaucracy. A sorrier bunch of ministers and generals than those gathered around the emperor is difficult to imagine. Already one could discern the outline of the catastrophe that would be triggered by Austrian chauvinism and lack of political judgment, in combination with Wilhelminian hubris and the militarism of the German Empire. Yet what unbelievably fertile soil for human culture was offered by this old, decaying monarchy in the midst of Europe! One never ceases to be surprised by the wealth of art, music, literature, and science during this era of political decline.

Here in Vienna, at the very heart of the old Europe, Erwin Chargaff was born into a fairly well-to-do Jewish family in 1905. His father had had a small bank, which he had lost together with his fortune before World War I. Nevertheless, he was lucky in that he died before the Nazis overran Austria. Chargaff's mother, on the other hand, of whom he paints a tender portrait in his autobiography, lived until 1943, when she disappeared without a trace in one of the Nazi *Nacht und Nebel* camps. Chargaff himself moved to the United States in the middle of the 1930s, for he had obtained a position as biochemist at Columbia University.

Chargaff is a fascinating and versatile man with deep interest in the humanities and an intimate knowledge of the European cultural heritage that has left an indelible mark on his personality. He has a certain intellectual arrogance and is not one to forgive his opponents easily. His incomparable scientific contribution is the discovery of base complementarity in DNA, a phenomenon we have already discussed. Chargaff, though, could never explain why DNA showed this base complementarity; his failure here is the tragedy of his scientific career. But no one can take from him the honor of having discovered the phenomenon.

Chargaff was well aware that his rules of base complementarity must imply something crucial about DNA structure. When in 1952 he went to Cambridge and met Watson and Crick, he certainly impressed them with the importance of the phenomenon. From a scientific point of view, and in terms of the consequences for elucidation of DNA structure, their meeting was a huge success. In every other way it was an unmitigated catastrophe. Chargaff seems to have taken an instant dislike to these unknown young men, and his low opinion of them was not helped by the fact that during their conversation they showed themselves to be shockingly ignorant of the detailed structure of the nucleotide building blocks in DNA. At least to Chargaff, with his scientific upbringing in the strict German tradition, this deficiency was unpardonable. It was comparable to someone venturing to discuss pyrimidines and purines with Emil Fischer without knowing their exact structure!

Chargaff had published his results in 1950, but it is doubtful that Watson and Crick were aware of his complementarity rules: by their own admission they were not so well acquainted with DNA chemistry. In his very readable and interesting autobiography, *Heraclitean Fire*, Chargaff claims to have put them on the right track during his stay in Cambridge and is obviously bitter that he has not been given full credit for that. He draws a picture of Crick and Watson that has the sharpness of real hatred. Crick is described as having

Felix Hoppe-Seyler (1825–1895)
Courtesy of the Royal Swedish Academy of Sciences, Stockholm.

August Weismann (1834–1914)
Courtesy of the Museum of Comparative Zoology, Harvard University, Cambridge, Massachusetts.

Albrecht Kossel (1853–1927)
Courtesy of the Nobel Foundation, Stockholm, Sweden.

Emil Fischer (1852–1919)
Courtesy of the Nobel Foundation, Stockholm, Sweden.

Adolf von Baeyer (1835–1917)
Courtesy of the Nobel Foundation, Stockholm, Sweden.

Phoebus Aaron Levene (1869–1940)
Courtesy of Jacques Fresco, Princeton University, Princeton, New Jersey.

Gregor Mendel (1822–1884)
Courtesy of the Royal Swedish Academy of Sciences, Stockholm.

Archibald Garrod (1857–1936)
Courtesy of the Royal Society, London.

Thomas Hunt Morgan (1866–1945)
Courtesy of the Nobel Foundation, Stockholm, Sweden.

Oswald Theodore Avery (1877–1955)
Reproduced from René J. Dubos, *The Professor, the Institute and DNA*, 1976, p. ii; by permission of Rockefeller University Press, New York.

Colin MacLeod (1909–1972) and Maclyn McCarty (1911–)
Reproduced from René J. Dubos, *The Professor, the Institute and DNA*, 1976, facing p. 129; by permission of Rockefeller University Press, New York.

James Watson (1928–)
Courtesy of the Nobel Foundation, Stockholm, Sweden.

Francis Crick (1916–   )
Courtesy of the Nobel Foundation, Stockholm, Sweden.

Erwin Chargaff (1905–)
Courtesy of Erwin Chargaff. Reproduced from Chargaff, *Heraclitean Fire*, 1978, jacket photo; by permission of Rockefeller University Press, New York.

Francis Crick (1916–)
Courtesy of the Nobel Foundation, Stockholm, Sweden.

Erwin Chargaff (1905–   )
Courtesy of Erwin Chargaff. Reproduced from Chargaff, *Heraclitean Fire,* 1978, jacket photo; by permission of Rockefeller University Press, New York.

Rosalind Franklin (1920–1958)
Courtesy of Jenifer Glynn, Daylesford, England.

Arthur Kornberg (1918–   ) with his wife, Sylvy, and their sons, Roger, Tom, and Ken
Courtesy of Arthur Kornberg, Stanford University, Stanford, California.

"the looks of a fading racing tout, something out of Hogarth" or as being a caricature by Daumier, in whose turbid stream of prattle occasional nuggets may be discerned. Watson is dismissed as "quite undeveloped at twenty-three, a grin, more sly than sheepish; saying little, nothing of consequence; a gawky young figure."

It is hard to recognize the two dominant figures of molecular genetics in Chargaff's spiteful picture, and it does not seem likely that he was the one who came up with the crucial idea of base pairs. After all, he admits he was never even close to the correct structural explanation of base complementarity. In any case, there can be no doubt that the triumphal progress around the world of the DNA double helix was a personal tragedy for Erwin Chargaff. Since then, his contributions to the general scientific debate have been increasingly critical. He has completely repudiated molecular genetics and its spectacular achievements, which he regards with contempt and mistrust and about whose real or imaginary risks he continually warns.

Chargaff provided Watson and Crick with a serious restriction in their model building, in that their DNA structure had to explain base complementarity. The great breakthrough came in the form of a brilliant idea: DNA is made up of two polynucleotide strands linked to each other by weak bonds, called hydrogen bonds. Such bonds occur when two atoms, donor and acceptor, share a hydrogen atom. A hydrogen bond cannot be formed unless donor, hydrogen, and acceptor are colinear (that is, lie on the same line in space) and the distances between them are optimal. Consequently, such bonds are fairly specific in terms of the molecular structures in which they can occur. In the case of DNA, it can be shown that the best hydrogen bonds are formed between A and T, and G and C. Assuming that the association between the putative two strands in DNA is dependent on such specific hydrogen bonds between complementary bases to form so-called base pairs, we have a perfect structural explanation of Chargaff's base complementarity. An A on one strand must correspond to a T on the other, and in the same way a G must correspond to a C. This idea of specific base pairing to hold the two DNA strands together is the very essence of Crick's and Watson's DNA structure. Once that decisive conceptual step had been taken, the rest was comparatively easy.

The idea of a helical configuration of the polynucleotide strands was suggested by Pauling's work, and with two such strands per DNA molecule the structure had to be a double helix. The problem was where to put the bases of the nucleotide building blocks. To begin with, Watson (who seems to have done most of the actual model building, while Crick's input was mainly intellectual) had always placed the bases on the outside of the double helix. The resulting structures, however, did not look reasonable. According to Crick's memoir, *What Mad Pursuit*, it was he who eventually persuaded Watson to place the bases on the inside. With that, everything fell into place. The complementary bases could interact to form the necessary hydrogen bonds, and bases could be stacked on top of one another to give sufficient stability to the DNA structure (Figure 3).

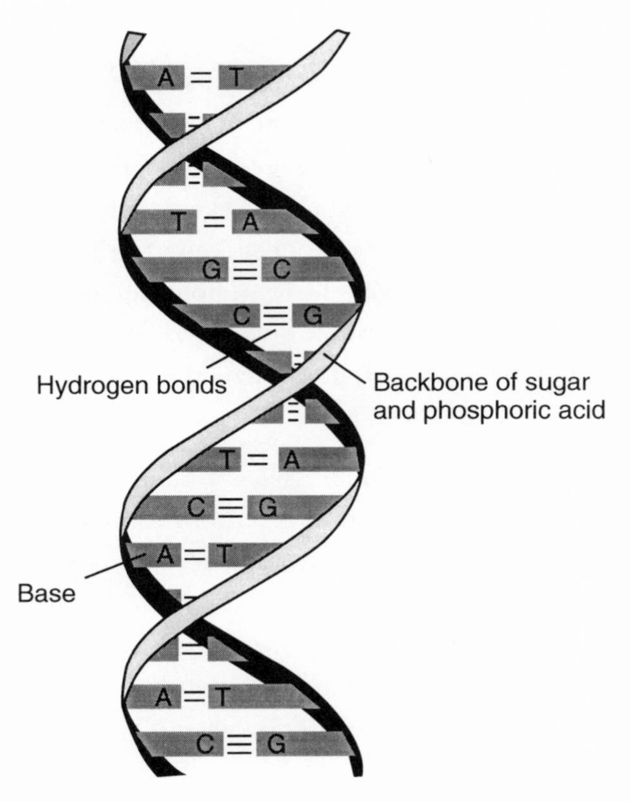

Figure 3 The double helix. Double-strand DNA consists of two chains of deoxynucleotides (strands) that are twisted around each other to form a double helix. The backbones of the strands are made up of sugar and phosphoric acid and delimit the double helix outward, while the bases are in the middle and thus can be stacked on each other, which adds greatly to the stability of the structure. The double helix is held together by the hydrogen bonds between the components of the specific base pairs G-C and A-T.

At last they had a model that looked valid in every respect. It must have been a moment of triumph. Crick tells us that he rushed home to his wife, Odile, and in his somewhat theatrical way told her about the unbelievably important discovery that he and Watson had just made. Many years later, Odile confided to him that she had not believed a word of his story, since from her point of view he came home all the time with fantastic yarns about incredible discoveries. Her skeptical attitude toward new and sensational hypotheses was in the best tradition of scientific research.

The double helix, as impressive an intellectual achievement as it was, at the same time certainly needed experimental support. That would come from two scientists in London, Maurice Wilkins and Rosalind Franklin, who in almost every way were the antithesis of Watson and Crick—in terms of their personalities, their mutual relationship, and their approach to the DNA problem. Franklin and Wilkins worked at King's College, London, on X-ray diffraction analysis of DNA fibers. It is a method related to crystallography, but incapable of giving the same detailed information. Unfortunately, no love was lost between the two researchers, an inauspicious start for a scientific collaboration.

Rosalind Franklin was a very competent crystallographer, who after much opposition from her upper-middle-class Jewish family had finally got her own way and embarked on a research career. She has been dead now for many years, but a picture of her as a young woman shows a strong-featured, almost beautiful face that has a certain seriousness, perhaps even sadness, about it. She was not in the least inclined to fanciful speculations of the kind that can under the right circumstances lead to daring leaps forward in science, but that in most cases end in spectacular belly flops. Franklin preferred to work slowly and systematically, one step at a time, and her relations with the model builders in Cambridge were far from cordial. On the contrary, we have every reason to suppose that she regarded them as two clowns, whose antics were not to be taken seriously. It did not help that Wilkins, her superior, was a friend of Crick's and found her approach to the DNA structure too cautious and unimaginative. Intellectual exchanges between the two groups were exceedingly sparse. All attempts to interest Franklin in model building of DNA came to nothing because of her firm conviction that it would be completely meaningless unless one was in possession of all the experimental data.

According to Crick, it was only after he and Watson completed their DNA model that they got access to Rosalind Franklin's X-ray diffraction data and realized the extent to which they supported the model. Despite the agreement between model and diffraction data, critics doubted the correctness of the double helix and rightly pointed out that experimental support of the model was far from unambiguous. Not until more than thirty years later, when it had become possible to crystallize synthetic fragments of DNA and analyze them by X-ray crystallography, was there definitive proof of the double helix. Nevertheless, it had an astounding impact on the scientific community and became a roaring success when it was presented in an article in *Nature* in 1953. Consider the very long time it took before people realized the true importance of Avery's great discovery. What was it, then, that made acceptance of the double helix so swift and so complete?

On the whole, scientists are a critical lot and not easily charmed. Outbursts of unrestrained enthusiasm are uncommon. Why the exception for "the golden helix"? One reason is that the model is so informative; it has obvious implications for the biological function of DNA and for the way it is formed in the cell (DNA replication). In their inimitable way Crick and Watson say in the *Nature* paper, "It has not escaped our notice that the specific pairing we have postulated immediately suggests a possible copying mechanism for the genetic material." In a second paper that appeared a month later in *Nature*, they gave a more detailed account of their ideas about DNA replication. It was soon discovered that this process is indeed based on the principle of base pairing.

A significant element in the immediate success of the double helix is what Günther Stent, one of the leading gurus of molecular biology, refers to as "style." That would obviously include the flamboyant introduction of the model onto the scientific scene, and perhaps also Crick's somewhat overwhelming personality. Another decisive factor is surely timing. There can be no doubt that the double helix arrived at exactly the right time. In classical drama, the author—after having first got the threads of the plot into a frightful tangle, thereby creating unbearable excitement and suspense in the audience—was in the habit of letting a deity, a *deus ex machina*, descend onto the stage at the critical moment to solve all the problems and put everything right again. In the drama we have seen performed on the genetic stage from Mendel to Avery, interest becomes more and more focused on one of the actors, DNA, which increasingly appears as the main figure but whose true nature is still shrouded in mystery. Suddenly Crick and Watson come drifting down from the higher spheres of acumen and creative fantasy, reveal the features of the enigmatic actor, bow to the ovations of the audience, and later receive a Nobel Prize.

No one can deny that it was indeed a well-deserved prize. But on the brightly lit stage one seems to perceive two tragic figures off in a dark corner—Rosalind Franklin, who four years earlier had died of cancer at the age of thirty-eight, and Erwin Chargaff, who had been so close to the goal, without being able to take the last decisive step.

[six]

How the Cell Makes DNA

The enormous amount of information that the cell needs for its vital functions, and that is codified as nucleotide sequences in its DNA, must be transferred from mother cell to daughter cells at the time the cell divides. If we for the moment disregard the special conditions that pertain to the production of mature gametes, the daughter cells should have the same genome as the mother cell that gave rise to them. We therefore have a fidelity problem. The doubling of the DNA content in the mother cell, which precedes cell division—the DNA replication—must not result in any change in the genetic information that will eventually be handed over to the daughter cells. Even an apparently simple organism such as *E. coli* has a genome consisting of a circular DNA molecule with almost four million base pairs. It is a formidable problem indeed to replicate so much DNA without any mistakes in the nucleotide sequences of the synthesized chains, but the cell can do this. Obviously it must contain extremely sophisticated enzymatic machinery, which can make identical replicas of the DNA with such high fidelity that the mistakes are kept at an acceptably low level. How is this machinery constructed, and how does it function?

The answer is very much the story of one man, Arthur Kornberg, and his achievements; but it is also a family story, since it is to a large extent about his scientific family—the laboratory he has created and that still bears the mark of his unique personality. It is also a family story in the sense that two of his sons have made significant contributions to DNA research.

Kornberg's parents emigrated to America at the turn of the century from Galicia, which before World War I was part of Austria-Hungary and also a center of European anti-Semitism. It is an irony of fate that the anti-Semitism in Central and Eastern Europe has resulted in the greatest brain drain that the Continent has suffered since Isabella and Ferdinand expelled the Moors and the Jews from Spain at the end of the fifteenth century. The exodus of European Jews to the United States has been of enormous importance for the rapid growth of American science. Nowhere is this more evident than in the fields of biochemistry and molecular biology. At the world famous Department of Biochemistry at Stanford University, for example, Arthur Kornberg has gathered a group of scientists who in many ways are unique. It is not only the accumulation of outstanding talents—that can be found in many comparable departments at top American universities—but it is the intense intellectual interactions that the members of the department have with one another. To this may be added the feeling of being in the midst of an incredibly gifted family, whose members seem amazingly to get on very well, something that is not always the case when outstanding scientists are crowded together. And their limited space was *terribly* crowded, at least until the department moved to new premises some years ago.

The road to Stanford was a long one for a poor Jewish boy from Brooklyn. His parents had to stretch their finances to give him the opportunity to study that they had never had. He went to school in New York, then studied medicine at the University of Rochester. There the leading light of the medical faculty, George H. Wipple, often expressed anti-Semitic feelings that Kornberg, more than fifty years later, recalled with considerable bitterness in his autobiography, *For the Love of Enzymes.* During World War II he joined the National Institutes of Health, at the time a rather modest establishment in Bethesda, Maryland, north of Washington. Today it has grown into a veritable anthill of scientists and is one of the world's leading centers of biomedical research. At the NIH Kornberg began a series of investigations into the biosynthesis of nucleotides that would make him internationally famous and lead to a chair in microbiology at Washington University in Saint Louis, where the brightest scientific stars were the biochemists Carl and Gerty Cori. By then Kornberg had become interested in the problem of DNA replication and Watson and Crick had just published their double-helix structure of DNA.

Scientists can be roughly divided into two main categories, slalom skiers and badgers. Slalom skiers move rapidly back and forth over the surface of the scientific field: in this way they can cover extensive areas in a comparatively short time and demonstrate strong intellectual versatility. Impressive as their activity may seem, it is somewhat superficial. The badgers, on the other hand, delve deep and continue digging until they have thoroughly explored the problem that they have taken on. There can be no doubt in what category Arthur Kornberg belongs. For more than forty years he has worked on DNA replication, and during all that time he has been the leading scientist in that intensely competitive field.

When Kornberg decided to tackle the problem of DNA replication, he chose *E. coli* as his experimental system. He started his investigations with a crude extract of the bacterium that should have contained all the enzymes necessary for the process. Using radioactively labeled deoxynucleosides, he tried to demonstrate the incorporation of those precursors of the building blocks of DNA, the deoxynucleotides (see Figure 1), into a DNA-like structure. The incorporation of isotopic material was barely detectable, but Kornberg and his collaborators continued doggedly and made steady progress. For instance, in order for the deoxynucleotides to be polymerized to DNA they had to be present as triphosphates, that is, in a form that gave them a special ability to transfer energy in chemical reactions. In addition to the deoxynucleoside-triphosphates corresponding to the deoxynucleotide constituents of DNA, the polymerizing enzyme (which Kornberg called DNA polymerase), required the presence of a DNA molecule to serve as a template in the reaction. In other

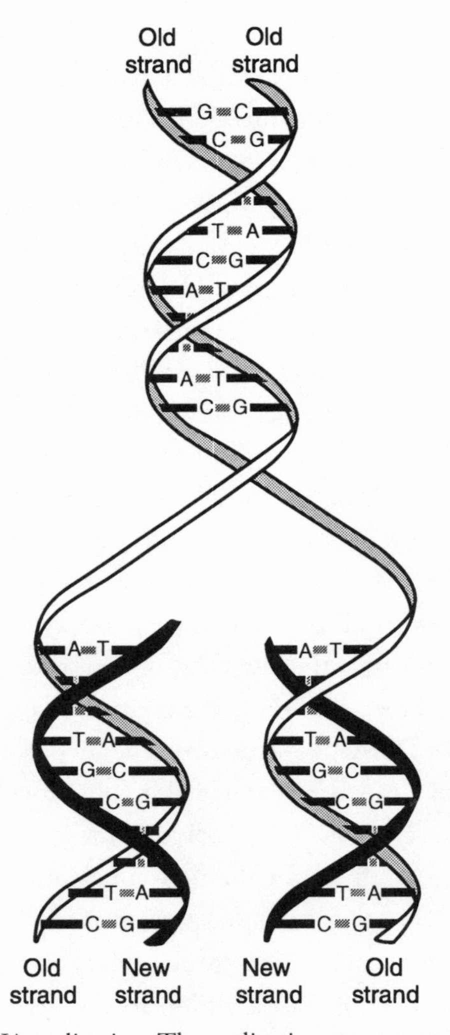

Figure 4 DNA replication. The replicative apparatus takes information from a DNA molecule that serves as a template and replicates it to form two exact copies. Both strands of the template are copied so that A in the template (old strand) corresponds to T in the new strand being synthesized, G corresponds to C, and so on. As a result, the new

words, the enzyme used the structural information of the template DNA to make what appeared to be a faithful copy of the DNA molecule in terms of its nucleotide sequence (Figure 4). At least, that was what it looked like to judge by the very primitive methods of analysis available at the time. Just as one would have predicted, DNA seemed to direct its own biosynthesis. This finding was highly encouraging and indicated that the Stanford group was on the right track.

After years of hard work, the DNA polymerase was obtained in a highly purified form and undoubtedly had many of the properties that could be expected of a DNA-replicating enzyme. In 1959 Arthur Kornberg shared the Nobel Prize in Physiology or Medicine with Severo Ochoa, in whose laboratory he had spent a happy year right after the war. At the same time he moved from Saint Louis, with its murderous climate, to Stanford.

When I arrived there for a sabbatical in 1960, I felt as though I was in the mecca of biochemistry. Science was the deity we all worshiped and Arthur Kornberg was its prophet—and for all this we had DNA polymerase to thank! Although I myself worked with one of Kornberg's colleagues, Paul Berg, on entirely different problems, I nevertheless felt that the glory of DNA polymerase reflected on everyone in

---

strand is complementary to the old strand being copied and can form a double helix with it, stabilized by the hydrogen bonds in the base pairs generated. A complete round of replication will result in two DNA molecules, each of which contains one old strand from the template and one new strand.

the laboratory. Certainly we all took great interest in the progress being made with that remarkable enzyme. But, alas, even the sun has its spots; DNA polymerase was found to be marred by some less-than-attractive features.

It was difficult to make the enzyme synthesize more than small quantities of DNA, compared to the amount of template DNA present in the incubation mixture. Furthermore, the DNA synthesized looked very weird in the electron microscope and seemed to have branching points, something never seen in native DNA. More worrisome still was the fact that the DNA produced by the enzyme could not be proved to have genetic activity. For instance, the DNA polymerase could not make viral DNA capable of infecting a suitable host, whereas native viral DNA can easily be shown to be infectious. It began to look as if the enzyme produced a DNA-like structure that lacked the biological activity of the template.

Could it be that the DNA polymerase was not the true DNA-replicating enzyme—that its real function in the cell was to repair damaged DNA? Such an assumption would agree with the early observation that the polymerase reaction was greatly facilitated if one "nicked" the template DNA with DNA-cleaving enzymes before incubating it with the DNA polymerase. A frustrating possibility indeed, but Arthur Kornberg was not one to give up easily. Perhaps things would improve if one allowed the DNA polymerase to copy a single-strand, circular DNA, where its task would be further simplified by the fact that the template contained no free end points.

We all know, of course, that DNA is a double helix, as originally suggested by Watson and Crick. But we also know that in nature there are exceptions to every rule. It is a matter of looking hard enough to find those exceptions, and that is true also of double-strand DNA. The outstanding biophysicist Robert Sinsheimer at the California Institute of Technology had for some time been interested in a bacteriophage (a virus that can infect bacteria) with the name ϕ X 174. Originally he had been fascinated by its remarkable smallness compared to other phages. He soon found that it had other, even more unusual properties. Its genome consisted of a circular, single-strand DNA molecule containing about five thousand nucleotides. When the phage infected its host, *E. coli*, that single-strand DNA was transformed into a double-strand, circular DNA molecule, the so-called replicative form, which could give rise to a number of copies through replication in the host cell. Why not try to copy this reaction in the test tube with the help of DNA polymerase?

In collaboration with Sinsheimer, Kornberg could indeed show that DNA polymerase was able to copy single-strand ϕ X 174 DNA to give a double-strand DNA. The problem was that DNA polymerase could not join the end points of the synthesized DNA chain to a circular molecule. Fortunately, at about this time several researchers, among them one of Kornberg's colleagues, Robert Lehman, had discovered an enzyme called DNA ligase, which had just the property that DNA polymerase so conspicuously lacked. Using these two enzymes in combination, Kornberg and his collaborators showed in 1967 that enzymatically synthesized, circular, single-strand DNA, corresponding to that found in the phage particle, could infect *E. coli* and thus had biological activity.

This was an exciting moment indeed. The public relations organization that exists at Stanford, as at all other major American universities, rose to the occasion and beat the drum. Even the White House felt repercussions of the discovery: President Lyndon Johnson himself stepped forward and said with characteristic modesty that American scientists had succeeded in producing life in the test tube, obviously as a result of the creative atmosphere provided by the Johnson administration. That was hardly what Arthur Kornberg had hoped to hear; rather, it confirmed his worst fears.

After all these successes, just as the pieces of the puzzle seemed to have fallen into place, a terrible anticlimax occurred. The English geneticist John Cairns reported in 1969 that he had isolated a mutant of *E. coli* that lacked DNA polymerase. It appeared to be healthy enough except that it had difficulty repairing DNA that had been damaged by ultraviolet radiation.

The report hit the DNA field like a bombshell. DNA polymerase after all had nothing to do with DNA replication; no organism could live without its replicative apparatus! A mutation that led to the loss of a replication enzyme had to be lethal, but Cairns's mutant was reasonably healthy. The unavoidable conclusion was that Kornberg's DNA polymerase was merely a repair enzyme.

All the jealousy that had lain dormant in the scientific community suddenly blossomed the world over, and the blooms did not smell like roses. *Nature*, that venerable patriarch among scientific journals, published a series of unpleasant editorials that obviously delighted in cutting Arthur Kornberg down to size. All the adverse publicity must have been depressing reading, but in the moment of misfortune Kornberg showed what he was made of. He lost no time moping, but immediately tackled the new problems that Cairns's mutant had revealed. Interestingly enough, one of the scientists who made an important contribution was Arthur's own son Tom.

To me Kornberg has always been the incarnation of the American family man. The support of his close-knit family, in particular of his wife, Sylvy, also a trained biochemist who worked with him in the laboratory for many years, has been of the utmost importance to him, both as a private person and as a scientist. It is now some thirty-five years since I first met his three sons at the pool of the Kornberg residence. They looked like very ordinary boys; but since then, Roger has made an international reputation for himself through his research on chromatin and transcription, while Tom has become a major name in DNA replication and developmental biology. Only Ken is quite normal, having chosen a more traditional career as an architect.

Tom early on showed signs of strong musical talent and was accepted at the Julliard School in New York, where he trained to become a cellist. Simultaneously he was a full-time student at Columbia College, where he took courses in biology and chemistry, proving that he was his father's son. Unfortunately, he fell victim to an occupational hazard of cellists: highly painful nerve-end tumors of his left index finger. He was forced to give up the cello and instead concentrated on biochemistry. Tom started to work with Malcolm Gefter at Columbia University and in a surprisingly short time succeeded in isolating two new DNA polymerases that he called DNA polymerase II and III (according to the generally accepted nomenclature, Arthur's original enzyme is designated DNA polymerase I). It is now clear that Tom's DNA polymerase III is the actual replicative enzyme, which has a very complex structure made up of a number of so-called subunits. This intricate and still not completely understood replicative enzyme is the main actor in the drama that takes place every time the cell replicates its DNA. As we shall see, there are many other important actors too.

When Arthur Kornberg started to reexamine the problems of DNA replication, he discovered that none of the DNA polymerases, I, II, or III, could start the synthesis of a new DNA chain. They could only elongate an existing chain—what we call a primer. His previous positive results with the replication of ϕ X 174 DNA were due to the presence of small fragments of *E. coli* DNA that served as primers in the incubation mixture. Some other enzyme in the cell had to be able to take directions from DNA and start the synthesis of a new chain of nucleotides, which could then be taken over by a DNA polymerase. Arthur determined that this useful helper was RNA primase, a special kind of RNA polymerase—an enzyme that copies DNA to produce RNA and had been discovered some ten years earlier. In other words, every DNA chain being synthesized starts with a short RNA sequence synthesized by RNA primase (Figure 5).

Another problem was that the two strands in the DNA molecule could not be replicated in the same way. One of the chains grows continuously during replication but the other chain, because of a structural feature of double-strand DNA that has to do with the orientation of the chains, must be replicated discontinuously to yield short DNA fragments that are later joined to each other. The Japanese biochemist Reiji Okazaki, a former collaborator of Arthur Kornberg's, had shown in 1967 that DNA replication gives rise to such fragments and that they are intermediates in the replication process. This discovery made him world famous but unfortunately he did not enjoy his fame for very long; he died of leukemia at the age of forty-five. As a boy, he had entered the ruins of Hiroshima after it was hit by the first atomic bomb, to look for his parents who had disappeared in the blast. Statistics seem to show that the frequency of leukemia is not higher in the surviving victims of the explosion than in the general population, but who can tell?

Each Okazaki fragment must originally start with a short RNA sequence that continues as a DNA chain. But mature DNA, which can be isolated from the cell, contains no RNA! There must be a mechanism for removing the RNA sequences and replacing them with DNA. This is where the old DNA polymerase I shows up again. It has an important role in replication, one that is very different from what we originally believed. Its role has to do with the basic function of the

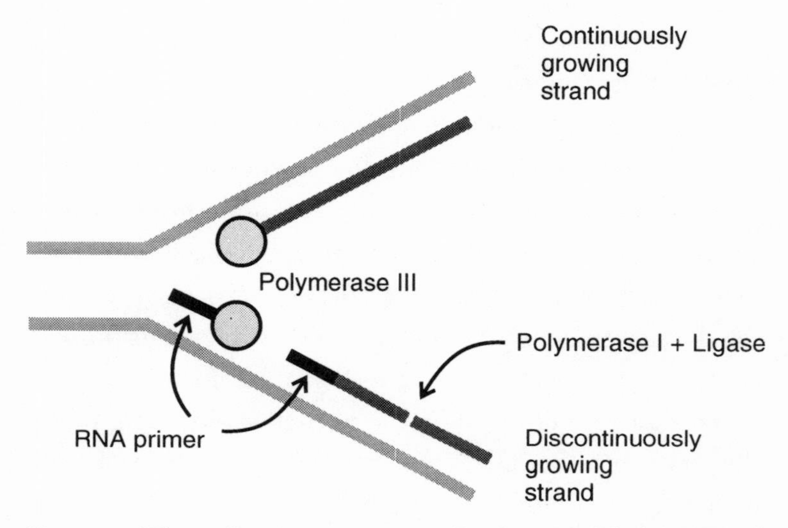

Figure 5 The replicative apparatus in *E. coli:* a simplified version of DNA replication. One of the new strands must, for structural reasons, grow discontinuously; that is, it is formed as short sequences (Okazaki fragments) that are later joined. The replicating enzyme, DNA polymerase III, like all other DNA polymerases, cannot start a new chain. It can only elongate an existing chain, a so-called primer. DNA polymerase III therefore requires the assistance of another enzyme, a specialized RNA polymerase called RNA primase. This enzyme can initiate synthesis and copy a short sequence of DNA to form an RNA primer, which can then be elongated by DNA polymerase III. Eventually the RNA sequences are removed, leaving defects in the DNA that are repaired by DNA polymerase I. Finally, the short DNA sequences—the Okazaki fragments—are joined together by DNA ligase to form a continuous DNA strand. Today more than a score of macromolecular components have been identified in DNA replication in *E. coli.*

enzyme, to repair DNA. In order to join the Okazaki fragments to form a continuous chain, one needs DNA ligase. Thus, both the enzymes, DNA polymerase I and DNA ligase, that were used in Arthur's famous ϕ X 174 experiment are in fact part of the DNA replication process. Everything has just turned out to be incredibly more complicated than our early suppositions. Today we have identified a score of different macromolecules that are needed for DNA replication in *E. coli.* Figure 5 gives a much-simplified version of our current view of this process.

Ever since he started to look into DNA replication more than forty years ago, Arthur Kornberg has been at the heart of this research. In recent years he has provided new and exciting information about the intricate mechanisms that direct and control the start of replication. We therefore have a much clearer picture of what happens when the cell initiates a replication cycle in order to be able to enter into the process of cell division.

Undoubtedly many admirers of Kornberg all over the world would subscribe to the view that "King Arthur" is our greatest living biochemist. One might ask what kind of person he is and what has made him so incredibly successful. To begin with, he has an extremely strong personality. Henry Kissinger tells us in his memoirs about his first meeting with Charles de Gaulle: he had the feeling that everything and everybody in the room, because of some mysterious quirk of the laws of nature, seemed to gravitate toward the place where the general was. One experiences something of the same feeling when Arthur enters the room.

Arthur Kornberg can certainly make a slightly overwhelming and perhaps intimidating first impression, but he has a large store of human warmth and concern under that image of magnificence. An example will illustrate. The wife of a young colleague fell victim to one of those mysterious psychotic episodes that sometimes occur after childbirth. One day she was discovered wandering around the streets in a completely confused condition. Guess who immediately took the matter in hand, tried to comfort her and calm her, then saw to it that she was admitted to the psychiatric clinic and given the best possible care? I have met quite a few great men in my life, but I do not think many of them would have done all that for someone they scarcely knew. To me at least, this says a lot about Arthur Kornberg's human qualities.

In his memoir, *For the Love of Enzymes*, Kornberg attempts to take stock of his life as a scientist. He asks himself what, of all he has accomplished, he feels is most important. One might predict he would select one of his major discoveries. Not so. It is the creation of the biochemistry department at Stanford, and the maintaining of its unique traditions, that he considers his most significant achievement. Perhaps he is right. In any case, it should be a comfort and an encouragement to all of us who have never made any outstanding discoveries of our own but who have striven to the best of our ability to deliver an important inheritance from one generation of scientists to the next.

Bibliography

Altmann, Richard. "Über Nucleinsäuren." *Müllers Archiv für Anatomie, Physiologie, und wissenschaftliche Medizin* (1889): 524–536.

Avery, Oswald T., Colin M. MacLeod, and Maclyn McCarty. "Studies on the Chemical Nature of the Substance Inducing Transformation of Pneumococcal Types." *Journal of Experimental Medicine* 79 (1944): 137–158.

Beadle, George W., and Edward L. Tatum. "Genetic Control of Biochemical Reactions in Neurospora." *Proceedings of the National Academy of Sciences (U.S.)* 27 (1941): 499–506.

Beneden, Edouard J. L. M. van. "L'appareil sexuel femelle de l'ascaride mégalocéphale." *Archives de biologie* 4 (1883): 95–142.

Bischoff, Theodor L. W. "Theorie der Befruchtung und über die Rolle, welche die Spermatozoiden dabei spielen." *Müllers Archiv für Anatomie, Physiologie, und wissenschaftliche Medizin* (1847): 422–442.

Chargaff, Erwin. "Chemical Specificity of Nucleic Acids and Mechanism of Their Enzymatic Degradation." *Experientia* 6 (1950): 201–209.

——— *Heraclitean Fire: Sketches from a Life before Nature*. New York: Rockefeller University Press, 1978.

Crick, Francis. *What Mad Pursuit: A Personal View of Scientific Discovery*. New York: Basic Books, 1988.

Dubos, René J. *The Professor, the Institute, and DNA*. New York: Rockefeller University Press, 1976.

Edlbacher, S. "Albrecht Kossel zum Gedächtnis." *Hoppe-Seyler's Zeitschrift für physiologische Chemie* 177 (1928): 1–14.

Fischer, Emil. "Untersuchungen über Aminosäuren, Polypeptide, und Proteine." *Berichte der Deutschen chemischen Gesellschaft* 39 (1906): 530–610.

——— *Aus meinem Leben*. Berlin: Julius Springer Verlag, 1922.

Flemming, Walther. *Zellsubstanz, Kern und Zellteilung*. Leipzig: Vogel, 1882.

Forster, Martin Onslow. "Emil Fischer Memorial Lecture." *Journal of the Chemical Society* 117 (1920): 1157–1201.

Fruton, Joseph S. *Molecules and Life*. New York: Wiley-Interscience, 1972.

Garrod, Archibald E. *Inborn Errors of Metabolism*. London: Frowde, Hodder, and Stoughton, 1909.

Goulian, Mehran, Arthur Kornberg, and Robert L. Sinsheimer. "Enzymatic Synthesis of DNA. XXIV. Synthesis of Infectious Phage ϕ X 174 DNA." *Proceedings of the National Academy of Sciences (U.S.)* 58 (1967): 2321–28.

Greenstein, Jesse P. "Friedrich Miescher, 1844–1895: Founder of Nuclear Chemistry." *Scientific Monthly* 57 (1943): 523–532.

Griffith, Fred. "The Significance of Pneumococcal Types." *Journal of Hygiene* 27 (1928): 113–159.

Hammarsten, Olof. "Zur Kenntnis der Nukleoproteide." *Hoppe-Seyler's Zeitschrift für physiologische Chemie* 19 (1894): 19–37.

Hershey, Alfred D., and Martha Chase. "Independent Functions of Viral Protein and Nucleic Acid in Growth of Bacteriophage." *Journal of General Physiology* 36 (1952): 39–56.

Hertwig, W. A. Oskar. "Das Problem der Befruchtung und der Isotropie des Eies, eine Theorie der Vererbung." *Jenaische Zeitschrift für Medizin und Naturwissenschaft* 18 (1885): 276–318.

His, Wilhelm. *Unsere Körperform und das physiologische Problem ihrer Entstehung*. Leipzig: Vogel, 1874.

Jaquet, A. "Professor Friedrich Miescher–Nachruf." *Verhandlungen der Naturforschenden Gesellschaft in Basel* 11 (1897): 399–417.

Jones, Mary Ellen. "Albrecht Kossel: A Biographical Sketch." *Yale Journal of Biology and Medicine* 26 (1953): 80–97.

Judson, Horace F. *The Eighth Day of Creation*. New York: Simon and Schuster, 1979.

Kingzett, Charles. *Animal Chemistry*. London: Longmans and Green, 1878.

Kornberg, Arthur. *For the Love of Enzymes: The Odyssey of a Biochemist.* Cambridge, Mass.: Harvard University Press, 1989.

Kossel, Albrecht, and Albert Neumann. "Über das Thymin, ein Spaltungsprodukt der Nucleinsäure." *Berichte der Deutschen chemischen Gesellschaft* 26 (1893): 2753–56.

Lederberg, Joshua. "What the Double Helix Has Meant for Basic Biomedical Science: A Personal Commentary." *Journal of the American Medical Association* 269 (1993): 15–19.

Levene, Phoebus A., and Walter A. Jacobs. "On the Structure of Thymus Nucleic Acid." *Journal of Biological Chemistry* 12 (1912): 411–420.

Levene, Phoebus A., and Robert S. Tipson. "The Ring Structure of Thymidine." *Journal of Biological Chemistry* 109 (1935): 623–630.

McCarty, Maclyn. *The Transforming Principle: Discovering That Genes Are Made of DNA*, ed. Lewis Thomas. New York: Commonwealth Fund, 1985.

Mendel, Gregor. "Versuche über Pflanzen-hybriden." *Verhandlungen des Naturforschenden Vereines in Brünn* 4 (1866): 3–47.

Miescher, Friedrich. "Über die chemische Zusammensetzung der Eiterzellen." *Hoppe-Seyler's medicinisch-chemische Untersuchungen* 4 (1871): 441–460.

——— "Die Spermatozoen einiger Wirbeltiere." *Verhandlungen der Naturforschenden Gesellschaft in Basel* 6 (1874): 138–208.

Minchin, Edward A. "The Evolution of the Cell." *American Naturalist* 50 (1916): 5–39.

Morgan, Thomas H. *The Theory of the Gene.* New Haven: Yale University Press, 1928.

Olby, Robert. *The Path to the Double Helix.* Seattle: University of Washington Press, 1975.

Portugal, Franklin H., and Jack S. Cohen. *A Century of DNA: A History of the Discovery of the Structure and Function of the Genetic Substance.* Cambridge, Mass.: MIT Press, 1977.

Slyke, Donald D. van, and Walter A. Jacobs. "Phoebus Aaron Levene, 1869–1940." *Biographical Memoirs, National Academy of Sciences* 23 (1945): 75–126.

Staehelin, Matthys. "Friedrich Miescher, der Entdecker der Nukleinsäuren." *Basler Stadtbuch* (1962): 134–162.

Strasburger, Eduard A. "The Minute Structure of Cells in Relation to Heredity." In *Darwin and Modern Science*, ed. A. C. Seward, pp. 102–111. Cambridge: Cambridge University Press, 1909.

Suter, F. "Professor Friedrich Miescher: Persönlichkeit und Lehrer." *Helvetica Physiologica Acta* (1944), Suppl. 2.

Thudichum, Johann L. W. "On Modern Text-books as Impediments to the Progress of Animal Chemistry." *Annals of Chemical Medicine* 2 (1881): 183–189.

Watson, James D. *The Double Helix.* London: Weidenfeld and Nicolson, 1968.

——— and Francis Crick. "A Structure for Deoxyribose Nucleic Acid." *Nature* 171 (1953): 737–738.

Wilson, Edmund B. *The Cell in Development and Inheritance.* New York: Macmillan, 1896.

Wolfram, Melville L. "Phoebus Aaron Levene." In *Great Chemists*, ed. Eduard Farber, pp. 1313–24. New York: 1961.

Worm-Müller, Jacob. "Zur Kenntnis der Nucleine." *Pflügers Archiv.* 8 (1874): 190–194.

Index